EFFECTIVE STRATEGIES AGAINST TERRORISM

HEARING

BEFORE THE

SUBCOMMITTEE ON NATIONAL SECURITY, EMERGING THREATS AND INTERNATIONAL RELATIONS

OF THE

COMMITTEE ON GOVERNMENT REFORM

HOUSE OF REPRESENTATIVES

ONE HUNDRED EIGHTH CONGRESS

SECOND SESSION

FEBRUARY 3, 2004

Serial No. 108–150

Printed for the use of the Committee on Government Reform

Available via the World Wide Web: http://www.gpo.gov/congress/house
http://www.house.gov/reform

U.S. GOVERNMENT PRINTING OFFICE

94–017 PDF WASHINGTON : 2004

For sale by the Superintendent of Documents, U.S. Government Printing Office
Internet: bookstore.gpo.gov Phone: toll free (866) 512–1800; DC area (202) 512–1800
Fax: (202) 512–2250 Mail: Stop SSOP, Washington, DC 20402–0001

COMMITTEE ON GOVERNMENT REFORM

TOM DAVIS, Virginia, *Chairman*

DAN BURTON, Indiana
CHRISTOPHER SHAYS, Connecticut
ILEANA ROS-LEHTINEN, Florida
JOHN M. McHUGH, New York
JOHN L. MICA, Florida
MARK E. SOUDER, Indiana
STEVEN C. LATOURETTE, Ohio
DOUG OSE, California
RON LEWIS, Kentucky
JO ANN DAVIS, Virginia
TODD RUSSELL PLATTS, Pennsylvania
CHRIS CANNON, Utah
ADAM H. PUTNAM, Florida
EDWARD L. SCHROCK, Virginia
JOHN J. DUNCAN, Jr., Tennessee
NATHAN DEAL, Georgia
CANDICE S. MILLER, Michigan
TIM MURPHY, Pennsylvania
MICHAEL R. TURNER, Ohio
JOHN R. CARTER, Texas
MARSHA BLACKBURN, Tennessee
——— ———
——— ———

HENRY A. WAXMAN, California
TOM LANTOS, California
MAJOR R. OWENS, New York
EDOLPHUS TOWNS, New York
PAUL E. KANJORSKI, Pennsylvania
CAROLYN B. MALONEY, New York
ELIJAH E. CUMMINGS, Maryland
DENNIS J. KUCINICH, Ohio
DANNY K. DAVIS, Illinois
JOHN F. TIERNEY, Massachusetts
WM. LACY CLAY, Missouri
DIANE E. WATSON, California
STEPHEN F. LYNCH, Massachusetts
CHRIS VAN HOLLEN, Maryland
LINDA T. SANCHEZ, California
C.A. "DUTCH" RUPPERSBERGER, Maryland
ELEANOR HOLMES NORTON, District of Columbia
JIM COOPER, Tennessee
——— ———

———

BERNARD SANDERS, Vermont (Independent)

MELISSA WOJCIAK, *Staff Director*
DAVID MARIN, *Deputy Staff Director/Communications Director*
ROB BORDEN, *Parliamentarian*
TERESA AUSTIN, *Chief Clerk*
PHIL BARNETT, *Minority Chief of Staff/Chief Counsel*

SUBCOMMITTEE ON NATIONAL SECURITY, EMERGING THREATS AND INTERNATIONAL RELATIONS

CHRISTOPHER SHAYS, Connecticut, *Chairman*

MICHAEL R. TURNER, Ohio
DAN BURTON, Indiana
STEVEN C. LATOURETTE, Ohio
RON LEWIS, Kentucky
TODD RUSSELL PLATTS, Pennsylvania
ADAM H. PUTNAM, Florida
EDWARD L. SCHROCK, Virginia
JOHN J. DUNCAN, Jr., Tennessee
TIM MURPHY, Pennsylvania
——— ———

DENNIS J. KUCINICH, Ohio
TOM LANTOS, California
BERNARD SANDERS, Vermont
STEPHEN F. LYNCH, Massachusetts
CAROLYN B. MALONEY, New York
LINDA T. SANCHEZ, California
C.A. "DUTCH" RUPPERSBERGER, Maryland
JOHN F. TIERNEY, Massachusetts
——— ———

EX OFFICIO

TOM DAVIS, Virginia
HENRY A. WAXMAN, California

LAWRENCE J. HALLORAN, *Staff Director and Counsel*
R. NICHOLAS PALARINO, *Senior Policy Advisor*
ROBERT A. BRIGGS, *Clerk*
ANDREW SU, *Minority Professional Staff Member*

CONTENTS

EFFECTIVE STRATEGIES AGAINST TERRORISM

TUESDAY, FEBRUARY 3, 2004

HOUSE OF REPRESENTATIVES,
SUBCOMMITTEE ON NATIONAL SECURITY, EMERGING
THREATS AND INTERNATIONAL RELATIONS,
COMMITTEE ON GOVERNMENT REFORM,
Washington, DC.

The subcommittee met, pursuant to notice, at 10:09 a.m., in room 2247, Rayburn House Office Building, Hon. Christopher Shays (chairman of the subcommittee) Presiding.

Present: Representatives Shays, Turner, Schrock, Murphy, Ruppersberger and Tierney.

Staff present: Lawrence Halloran, staff director and counsel; R. Nicholas Palarino, senior policy advisor; Thomas Costa, professional staff member; Robert A. Briggs, clerk; Andrew Su, minority professional staff member; and Jean Gosa, minority assistant clerk.

Mr. SHAYS. A quorum being present, the Subcommittee on National Security, Emerging Threats and International Relations hearing entitled, "Effective Strategies Against Terrorism," is called to order.

Scientists remind us the plural of anecdote is not data. In the realm of national security, a similar axiom would hold the proliferation of counterterrorism strategies does not necessarily mean we are any safer. Only if those strategies guide us inexorably and immeasurably toward clearly articulated goals will they secure our liberty and prosperity against the threats of a new and dangerous era.

Prior to September 11, 2001, this subcommittee heard testimony based on the work of the three national commissions on terrorism—Bremer, Gilmore and Hart-Rudman—citing the lack of any overarching counterterrorism strategy. Last year, witnesses told us the Bush administration had succeeded in filling the strategic void with no less than eight high-level mission statements on national security, military strategy, global terrorism, homeland security, weapons of mass destruction, money laundering, cybersecurity, and critical infrastructure.

These strategies suggest the need for a post-cold war security paradigm that replaces containment and mutually assured destruction with detection, prevention and, at times, preemptive action to protect the fundamental interests of the United States. But the multi-dimensional threat of terrorism demands levels of strategic dynamism, flexibility and accountability never required to meet the relatively static Soviet menace. So we asked the General Accounting Office [GAO], to describe the fundamental characteristics of a coherent framework; one that clearly states a purpose, assesses risk, sets goals, defines needed resources, assigns responsibilities, and integrates implementation.

According to their analysis, current strategies contain many of these traits to some degree, but do not yet include key elements, particularly in the area of resource implementation and coordination to avoid duplication.

Yesterday, the President's proposed budget for the next fiscal year outlined the near and long-term costs of the war against terrorism. The strategies under discussion here today contain the words that are supposed to be driving those numbers toward achievement of higher level of tangible national goals. How can those strategies be clear, more concrete, and more tightly integrated into an inescapably logical whole? How will we know programs are achieving strategic objectives?

Testimony by GAO and by our second panel of expert witnesses will help us understand those questions and assess the strength and weaknesses of current counterterrorism strategies. We are very grateful for the insight and expertise they bring to our ongoing oversight, and we look forward to their testimony.

[The prepared statement of Hon. Christopher Shays follows:]

ONE HUNDRED EIGHTH CONGRESS

Congress of the United States

House of Representatives

COMMITTEE ON GOVERNMENT REFORM

2157 RAYBURN HOUSE OFFICE BUILDING

WASHINGTON, DC 20515–6143

MAJORITY (202) 225–5074
FACSIMILE (202) 225–3974
MINORITY (202) 225–5051
TTY (202) 225–6852

www.house.gov/reform

SUBCOMMITTEE ON NATIONAL SECURITY, EMERGING THREATS, AND INTERNATIONAL RELATIONS

Christopher Shays, Connecticut
Chairman
Room B-372 Rayburn Building
Washington, D.C. 20515
Tel: 202 225-2548
Fax 202 225-2382
E-mail: hr.groc@mail.house.gov

Statement of Rep. Christopher Shays
February 3, 2004

Scientists remind us the plural of "anecdote" is not "data." In the realm of national security, a similar axiom would hold the proliferation of counterterrorism strategies does not necessarily mean we are any safer. Only if those strategies guide us inexorably and measurably toward clearly articulated goals will they secure our liberty and prosperity against the threats of a new and dangerous era.

Prior to September 11th 2001, this Subcommittee heard testimony based on the work of the three national commissions on terrorism – Bremer, Gilmore and Hart-Rudman – citing the lack of any overarching counterterrorism strategy. Last year, witnesses told us the Bush Administration had succeeded in filling the strategic void with no less than eight high-level mission statements on: national security, military strategy, global terrorism, homeland security, weapons of mass destruction, money laundering, cyber security and critical infrastructure.

These strategies address the need for a post-Cold War security paradigm that replaces containment and mutually assured destruction with detection, prevention, and at times preemptive action to protect the fundamental interests of the United States. But the multi-dimensional threat of terrorism demands levels of strategic dynamism, flexibility and accountability never required to meet the relatively static Soviet menace.

So we asked the General Accounting Office (GAO), to describe the fundamental characteristics of a coherent strategic framework; one that clearly states a purpose, assesses risk, sets goals, defines needed resources, assigns responsibilities and integrates implementation. According to their analysis, current strategies contain many of these traits to some degree, but do not yet include key elements, particularly in the areas of resource implications and coordination to avoid duplication.

Yesterday, the President's proposed budget for the next fiscal year outlined the near- and long-terms costs of the war against terrorism. The strategies under discussion here today contain the words that are supposed to be driving those numbers toward achievement of high-level but tangible national goals. How can those strategies be clearer, more concrete and more tightly integrated into an inescapably logical whole? How will we know if programs are achieving strategic objectives?

Testimony by GAO, and by our second panel of expert witnesses, will help us answer these questions and assess the strengths and weaknesses of current counterterrorism strategies. We are grateful for the insight and expertise they bring to our ongoing oversight and we look forward to their testimony.

Mr. SHAYS. At this time, the Chair would recognize the vice chairman of the committee, the gentleman from Ohio.

Mr. TURNER. Mr. Chairman, thank you.

I just want to continue to appreciate your focus on these issues, and I look forward to the testimony today.

Mr. SHAYS. Thank you very much; and the gentleman from Virginia.

Mr. SCHROCK. Thank you, Mr. Chairman; and thank you for holding this hearing on a most important aspect of national security. It is indeed a fitting and appropriate way for us to begin this session.

I also want to thank all of the witnesses for lending their expertise to this committee's efforts to better understand and evaluate this matter. That the events of September 11, 2001, in their scale and audacity were such an unexpected invasion upon our sense of safety and control of our lives and that a small number of terrorists could strike such a devastating blow gives a sense of urgency to our need to distill our security division.

The National Security Strategy put forth by this administration in September 2002, is a commendable step in this effort to focus our military law enforcement and diplomatic resources to enhancing our security.

Like many members of this committee I still have grave concerns about our ability to integrate the efforts working to make this country more secure, particularly with respect to intelligence gathering and sharing. I am confident that, given the urgency of the war on terror, we all feel that as a Nation we will continue to identify our weaknesses and work to improve and rise to the challenge.

Again, Mr. Chairman, thank you for holding this hearing to advance us toward this goal and to the witnesses for both their time in testifying and analyzing this important effort.

Thank you, Mr. Chairman.

Mr. SHAYS. Thank you, Mr. Schrock, and thank you as well for your really faithful participation on this committee. I'd like to align myself with your comments.

I'd ask unanimous consent that all members of the subcommittee be permitted to place an opening statement in the record, and without objection so ordered, and that the record remain open for 3 days for that purpose.

I ask further unanimous consent that all witnesses be permitted to include their written statements in the record, and without objection so ordered.

At this time, we will recognize our first panel, comprised of one individual, Mr. Randall Yim, Managing Director of Homeland Security and Justice Team, U.S. General Accounting Office.

Mr. Yim, if you will stand, we will swear you in and then begin the testimony.

[Witness sworn.]

Mr. SHAYS. Thank you.

We appreciate your presence here today and the terrific work that GAO does on so many issues. You and your colleagues are invaluable to the work of this committee and to the work of Congress.

With that, what we'll do is we have 5 minutes. We'll roll it over another 5 minutes.

Is the clock working? OK.

STATEMENT OF RANDALL A. YIM, MANAGING DIRECTOR, HOMELAND SECURITY AND JUSTICE TEAM, U.S. GENERAL ACCOUNTING OFFICE

Mr. YIM. Thank you, Mr. Chairman.

Mr. Chairman, Vice Chairman Turner, Ranking Member Kucinich, Mr. Schrock, members of the committee, thank you for providing GAO with this opportunity to contribute to our Homeland Security efforts.

We undertook this work at this committee's request to constructively assist the Congress and the executive agencies in moving our Nation forward, in sync, in concert, with the available resources in a balanced, measured, and measurable manner toward better Homeland Security and national preparedness.

We hope that our testimony today assists in the evolution and implementation of national strategies so that Homeland Security efforts nationwide are clear, sustainable, integrated into agency governmental and private sector missions, helps in the difficult decisions in balancing Homeland Security priorities with other national objectives and ensures transparency needed for effective oversight and accountability.

In our review, we recognize that the national strategies are only beginning starting points for other parties developing more detailed implementation plans; and we recognize that the true measure of these strategies will be determined through time as they are implemented by the Federal, State, local and private international sectors and as Homeland Security actions are embedded or integrated into ongoing governmental and private sector missions in sustainable, balanced ways.

Thus, the value of these strategies will be the extent to which they are useful for and actually used by the responsible parties to guide their own actions, to make difficult resourcing decisions and to develop and maintain their assigned capabilities to respond as expected when needed.

This means that the strategies must be relevant and useful not only during times of crisis but during prolonged times of preparedness. The strategies must be useful for all phases of our Homeland Security efforts, prevention, vulnerability assessment, reduction response and recovery; and these strategies should be used not just when an emergency arises, when there is a danger of panic driven activities, but during the hopefully increasingly long periods of time when there are no attacks, no horrific situations that consume our attention.

I recently spoke at a senior commanders' conference for the Joint Command that includes the military district of Washington. One of the concerns raised by the senior leaders is that we must act now to define and coordinate the responsibilities of the Federal, State and local governments and the private sector while their memories of September 11 are still in the forefront before complacency sets in and hampers our efforts.

Indeed, a survey of about 1,400 private CEOs presented at the World Economic Forum rates global terrorism only tied for 6th on the list of 11 challenges that these CEOs view to the biggest threat to their companies.

Our Nation must make the necessary steps to improve Homeland Security now with a sense of urgency. The strategies must make such improvements even without an immediate emergent situation.

What did we find?

We found that the national strategies are not required by executive or legislative mandate to address a single set of characteristics and, not surprisingly, they contain varying degrees of detail based upon their scopes and maturity in their underlying programs.

Further, we found that there is no commonly accepted set of characteristics used for a national strategy. As a result, after consulting with numerous sources, GAO developed a set of desirable characteristics that we believe are critical to provide effective guidance. These are: a statement of purpose scope and methodology; second, a problem of risk definition and assessment; third, identification of goals, subordinate objectives, activities, and performance measures; fourth, resource investment and risk management discussions; fifth, organizational roles responsibilities and coordination; and, finally, integration and implementation.

We then evaluated the seven national strategies by the extent to which they contain these key characteristics. The seven strategies we evaluated were: the National Security Strategy of the United States, September 2002, publication; the National Strategy for Homeland Security in July 2002; the National Strategy for Combating Terrorism in February 2003; the National Strategy to Combat Weapons of Mass Destruction in December 2002; the National Strategy for Physical Protection of Critical Infrastructure and Key Assets, February 2003; the National Strategy to Secure Cyberspace, February 2003; and the 2002 Money Laundering Strategy.

Page 4 of my testimony contains a matrix summarizing the results of our evaluation, and I'd like to emphasize certain points on that table. Five of these points are newly published in September 11 and relate to specific areas of homeland security and combating terrorism. The other two strategies, the National Security Strategy and the 2002 Money Laundering Strategy, were updated from pre-September 11 versions, and only these two strategies are required by statutes that mandate specific content elements.

Thus, admittedly the six identified key characteristics and the evaluation of the extent to which the strategies address these characteristics have a degree of subjectivity, even though we at GAO follow consistent and clear criteria during our evaluation.

Because of this inherent subjectivity, the value of our analysis lies not in an absolute or stand-alone assessment of the strategies. That is, we are not attempting to assign an absolute grade to the strategy but rather a comparative analysis between and among the strategy. Some are better in our views than others. Some employ best practices that have enhanced value to the users.

Our objective is to learn from the best to assist this Congress in continually evolving these strategies in an expedited matter.

The strategies generally do not address resourcing risk management and implementation. Those desired objectives are not clearly linked to funding and sustainability.

How are we going to pay for homeland security measures, who should pay, how do we factor in costs—effectiveness? How do we implement additional homeland security without consequences such as deleterious impacts upon businesses or civil liberties, privacy issues; and, second, even where the desirable characteristics are addressed, the strategies could be improved.

Of course, while strategies identify goals, subordinate objectives and specific activities, they generally do not discuss or identify priorities, milestones or performance measures that we consider are crucial to effective oversight and decisionmaking. So let me briefly touch upon those six characteristics with a specific example.

First, purpose, scope, and methodology. Fundamentally, a good strategy has to identify what it does and it does not cover so that the users know what to expect and the right people are brought together for both development and implementation.

Importantly, key definitions can provide the clarity necessary. For example, some of the earlier iterations of the critical infrastructure protection strategy defined it as cyberstrategy, as opposed to physical structures. That was clarified later, as to help the users agree upon a problem to be addressed in some means to determine priorities. So some strategies like money strategy focuses on law enforcement, others on deterrence, others on prevention and response; and that can sometimes lead to conflicts or tensions between the agencies because sometimes law enforcement is incompatible with crime scene response. So it's very difficult. We have to define problems, set priorities. We have to do it fundamentally on a risk basis by identifying threats, identifying vulnerabilities and the cascading impacts, should a threat come to fruition.

The Homeland Security Strategy does have a separate threat and vulnerability section, but many others do not.

Third one, goals, performance measures. Obviously, we would like to have a hierarchy of goals to achieve those end-states.

Performance or out-commissioned goals, as opposed to some of the mistakes we made in the Department of Defense of prescribing specific solutions, allow responsible parties to develop integrated approaches and to tailor it to specific sectors or regions; and they allow us some accountability both as to the use of funds but also are people capable of assuming assigned responsibilities once the strategies make those assignments.

Next category, resource investment and risk management. The strategy should address cost issues, how much, who's going to pay, how are we going to pay, the types of resources and investments associated. I think they all make the logical assumption that we cannot afford to do everything, so we have to have some rational risk management approach to do the things that are best within our available resources to stretch and leverage our resources. For example, the cyberspace strategy relies upon market-driven approaches because of rapid changing technology in that arena. However, on other sectors that don't move as quickly, bridges or transit, perhaps another strategy could be employed.

Organizational roles and responsibility is a fundamental question of who's in charge of not only during times of crisis but during what I said, times of prolonged preparedness.

Who's in charge. Also, let's us coordinate the activities among various responsible parties. The Money Laundering Strategy is a good example. It assigns specific objectives.

And, finally, integration and implementation. We will never be fully successful in our homeland security strategies if we continue to see homeland security as a separate cost activity. We will and should overlap with other national important strategies. We have to talk about designing in homeland security up-front at the same time we're talking about recapitalizing our infrastructure, rather than trying to retrofit our infrastructure; and I think that these types of integration will help us strike fundamental balances of the many important things our citizens are asking the government to do.

So where do we go from here? I'd like to conclude my oral comments with a few observations and suggestions.

As I said before, the ultimate test of the strategy will be determined through time as they're implemented. Are they useful? Are they actually being used by the parties responsible?

So it's going to be very responsible for GAO, this committee, the Congress, the administration, to solicit input from all responsible parties, State and local, international and incorporate this to ensure improved preparedness. The feedback will be to this committee, and obstacles will be identified that may require legislative action if necessary. Feedback to the Congress will also allow us to improve our grant systems and other stimulus and investment programs. Mechanisms that set performance metrics will really help us tell if we're getting our money's worth.

Finally, integration and implementation may be enhanced by national standards that link together these responsible parties using management and systems principles that are analogous to some of the very recognized ISO-type management standards that have been used.

Much has been done, Mr. Chairman; much more needs to be done; and GAO looks forward to working with this committee.

Thank you.

Mr. SHAYS. Thank you, Mr. Yim, for your testimony and for all your good work.

[The prepared statement of Mr. Yim follows:]

United States General Accounting Office

GAO

Subcommittee on National Security, Emerging Threats, and International Relations, Committee on Government Reform, House of Representatives

For Release on Delivery
Expected at 10:00 a.m. EST
Tuesday, February 3, 2004

COMBATING TERRORISM

Evaluation of Selected Characteristics in National Strategies Related to Terrorism

Statement of Randall A. Yim, Managing Director
Homeland Security and Justice Issues

GAO
Accountability * Integrity * Reliability

GAO-04-408T

February 3, 2004

GAO
Accountability · Integrity · Reliability

Highlights

Highlights of GAO-04-408T, testimony before the Subcommittee on National Security, Emerging Threats, and International Relations, Committee on Government Reform, House of Representatives

COMBATING TERRORISM

Evaluation of Selected Characteristics in National Strategies Related to Terrorism

Why GAO Did This Study

Following the terrorist attacks of September 11, 2001, the Bush administration developed and published seven national strategies that relate, in part or in whole, to combating terrorism and homeland security. These were the:

- *National Security Strategy of the United States of America.*
- *National Strategy for Homeland Security.*
- *National Strategy for Combating Terrorism.*
- *National Strategy to Combat Weapons of Mass Destruction.*
- *National Strategy for the Physical Protection of Critical Infrastructure and Key Assets.*
- *National Strategy to Secure Cyberspace.*
- *2002 National Money Laundering Strategy.*

In view of heightened concerns about terrorism and homeland security, GAO was asked to identify and define the desirable characteristics of an effective national strategy and to evaluate whether the national strategies related to terrorism address those characteristics. The purpose of this testimony is to report on GAO's findings on this matter.

www.gao.gov/cgi-bin/getrpt?GAO-04-408T.

To view the full product, including the scope and methodology, click on the link above. For more information, contact Randall A. Yim at (202) 512-6787 or YimR@gao.gov.

What GAO Found

National strategies are not required by either executive or legislative mandate to address a single, consistent set of characteristics. However, based on a review of numerous sources, GAO identified a set of desirable characteristics to aid responsible parties in further developing and implementing the strategies—and to enhance their usefulness in resource and policy decisions and to better assure accountability. The characteristics GAO identified are: (1) purpose, scope, and methodology; (2) problem definition and risk assessment; (3) goals, subordinate objectives, activities, and performance measures; (4) resources, investments, and risk management; (5) organizational roles, responsibilities, and coordination; and (6) integration and implementation.

GAO found considerable variation in the extent to which the seven strategies related to combating terrorism and homeland security address the desirable characteristics. A majority of the strategies at least partially address the six characteristics. However, none of the strategies addresses all of the elements of resources, investments, and risk management; or integration and implementation. Even where the characteristics are addressed, improvements could be made. For example, while the strategies identify goals, subordinate objectives, and specific activities, they generally do not discuss or identify priorities, milestones, or performance measures—elements that are desirable for evaluating progress and ensuring effective oversight. On the whole, the *National Strategy for Homeland Security* and the *National Strategy for the Physical Protection of Critical Infrastructure and Key Assets* address the greatest number of desirable characteristics, while the *National Security Strategy* and the *National Strategy to Combat Weapons of Mass Destruction* address the fewest.

The Pentagon in Flames Moments after International Terrorists Crash a Hijacked Aircraft into the Building on September 11, 2001

Source: U.S. Marine Corps.

Mr. Chairman and Members of the Subcommittee:

I appreciate the opportunity to be here today to participate in this hearing that examines the various national strategies published by the Bush Administration following the terrorist attacks of September 11, 2001. These strategies represent the administration's guidance to the federal state, local, private, and international sectors, for combating terrorism and securing the homeland and, equally important, for sustaining efforts into the future. Specifically, these seven strategies cover a broad range of related topics—from preparing against terrorist attacks to combating weapons of mass destruction, protecting our physical infrastructure, securing cyberspace, and blocking terrorist financing. The new strategies accompany the federal government's biggest reorganization in more than 50 years, resulting in the creation of a new Department of Homeland Security (DHS) to address the new threat environment.

Based upon heightened concerns about terrorism and homeland security, the Subcommittee asked us (1) to identify and define the characteristics of an effective national strategy and (2) to evaluate whether the strategies related to terrorism address those characteristics. This work expands upon our testimony to the Subcommittee in March 2003 and a related report in May 2003, as well as prior work for this Subcommittee and other committees over the past 7 years.[1]

After providing some background on the strategies related to terrorism, my statement will identify a set of desirable characteristics for any effective national strategy and compare and contrast the extent to which each of the strategies we address contains such characteristics. We believe these desirable characteristics would help shape the policies, programs, priorities, resource allocations, and standards that would enable federal agencies and other stakeholders to implement the strategies and achieve the identified results. We hope that the value of our review lies in assisting the evolution and implementation of these national strategies, so that homeland security efforts nationwide are clear, sustainable, and integrated into agency, governmental, and private sector missions; and, further, that

[1] See U.S. General Accounting Office, *Combating Terrorism: Observations on National Strategies Related to Terrorism*, GAO-03-519T (Washington, D.C.: Mar. 3, 2003) and *Combating Terrorism: Interagency Framework and Agency Programs to Address the Overseas Threat*, GAO-03-165 (Washington, D.C.: May 2003). In addition, a list of related GAO products is at the end of this statement.

these efforts are balanced with other important priorities, and transparent enough to ensure accountability.

We recognize the difficulty of the tasks presented to the strategy developers—and that national strategies are only starting points for federal agencies and other parties responsible for developing more detailed implementation plans. In some areas, so much needed to be done quickly that even general strategic statements added value. Some of the differences in detail in the national strategies may be attributed to their different breadths of scope and/or the maturity levels in their underlying program activities. We hope it is instructive to compare and contrast these strategies not only to each other, but also with other complex strategic planning efforts, so that the value of the strategies as guidance is enhanced and the timeframe for further refinements and implementation is expedited, given the critical nature of our homeland security efforts.

The new or updated national strategies released in the past 2 years that relate to combating terrorism and homeland security, in part or in whole, are:

- The *National Security Strategy of the United States of America*, September 2002.
- The *National Strategy for Homeland Security*, July 2002.
- The *National Strategy for Combating Terrorism*, February 2003.
- The *National Strategy to Combat Weapons of Mass Destruction*, December 2002.
- The *National Strategy for the Physical Protection of Critical Infrastructure and Key Assets*, February 2003.
- The *National Strategy to Secure Cyberspace*, February 2003.
- The *2002 National Money Laundering Strategy*, July 2002.

As agreed with your staff, we will report separately on the classified *National Military Strategic Plan for the War on Terrorism.*

Summary

National strategies are not required by executive or legislative mandate to address a single, consistent set of characteristics, and they contain varying degrees of detail based on their different scopes. Furthermore, we found there was no commonly accepted set of characteristics used for an effective national strategy. Nonetheless, after consulting numerous sources, we identified a set of desirable characteristics that we believe would provide additional guidance to responsible parties for developing and implementing the strategies—and to enhance their usefulness as guidance for resource and policy decision-makers and to better ensure accountability. Those characteristics are: (1) a statement of purpose, scope, and methodology; (2) problem definition and risk assessment; (3) goals, subordinate objectives, activities, and performance measures; (4) resources, investments, and risk management; (5) organizational roles, responsibilities, and coordination; and (6) integration and implementation. We identified these desirable characteristics by consulting statutory requirements pertaining to certain strategies we reviewed, as well as legislative and executive branch guidance for other national strategies. In addition, we studied the Government Performance and Results Act of 1993 (GPRA); general literature on strategic planning and performance; and guidance from the Office of Management and Budget (OMB) on the President's Management Agenda. We also gathered published recommendations made by national commissions chartered by Congress; past GAO work; and various research organizations that have commented on national strategies.

The seven national strategies related to homeland security and combating terrorism vary considerably in the extent to which they address the desirable characteristics that we identified. All seven strategies we reviewed partially address goals, subordinate objectives, activities, and performance measures. Four of the strategies address problem definition and risk assessment, while one strategy partially addresses that characteristic. And a majority of the strategies at least partially address the four other characteristics: purpose, scope, and methodology; resources, investments, and risk management; organizational roles, responsibilities, and coordination; and integration and implementation. However, none of the strategies addresses all of the elements of resources, investments, and risk management; or integration and implementation. Furthermore, even where the strategies address certain elements of the characteristics, there is room for improvement. For example, while the strategies identify goals, subordinate objectives, and specific activities, they generally do not discuss or identify priorities, milestones, or performance measures—elements that we consider to be desirable for evaluating progress, achieving results, and ensuring effective oversight. On the whole, the *National Strategy for Homeland Security* and the *National Strategy for the Physical Protection of Critical Infrastructure and Key Assets* address the greatest number of the desirable characteristics, while the *National Security Strategy* and the *National Strategy to Combat Weapons of Mass Destruction* address the fewest. Table 1 shows the extent that the strategies address, partially address, or do not address our characteristics.

Table 1: National Strategies and the Extent they Address GAO's Desirable Characteristics

National Strategy (short titles)	Purpose, scope, and methodology	Problem definition and risk assessment	Goals, subordinate objectives, activities, and performance measures	Resources, investments, and risk management	Organizational roles, responsibilities, and coordination	Integration and implementation
National Security	Does not address	Does not address	Partially addresses	Does not address	Does not address	Does not address
Homeland Security	Addresses	Addresses	Partially addresses	Partially addresses	Addresses	Partially addresses
Combating Terrorism	Partially addresses	Addresses	Partially addresses	Does not address	Partially addresses	Partially addresses
Weapons of Mass Destruction	Does not address	Does not address	Partially addresses	Does not address	Partially addresses	Partially addresses
Physical Infrastructure	Addresses	Addresses	Partially addresses	Partially addresses	Partially Addresses	Partially addresses
Secure Cyberspace	Partially addresses	Addresses	Partially addresses	Partially addresses	Partially Addresses	Partially addresses
Money Laundering	Partially addresses	Partially addresses	Partially addresses	Partially addresses	Partially addresses	Partially addresses

Source: GAO analysis.

Note: Per our methodology, a strategy "addresses," a characteristic when it explicitly cites all elements of a characteristic, even if it lacks specificity and details and thus could be improved upon. A strategy "partially addresses" a characteristic when it explicitly cites some, but not all elements of a characteristic. Within our designation of "partially addresses" there is a wide variation between a strategy that addresses most of the elements of a characteristic and a strategy that addresses few of the elements of a characteristic. A strategy "does not address" a characteristic when it does not explicitly cite or discuss any elements of a characteristic, and/or any implicit references are either too vague or general. See appendix I for more details on our methodology.

Background

Seven National Strategies Related to Combating Terrorism Released Since September 11 Attacks

In the wake of the terrorist attacks on September 11, 2001, seven new national strategies were developed and published to help guide U.S. efforts to combat terrorism. Of these, five were newly published strategies that related to specific aspects of homeland security and combating terrorism, such as weapons of mass destruction, protecting physical infrastructure, and securing cyberspace. Two strategies, the *National Security Strategy of the United States of America* and *the 2002 National Money Laundering Strategy*, were updated from pre-September 11 versions to specifically include terrorism. "Terrorism" may be generally defined as politically motivated violence to coerce a government or civilian population. "Combating terrorism" refers to the full range of policies, programs, and activities to counter terrorism, both at home and abroad. There is a further distinction within "combating terrorism," with "homeland security" referring to domestic efforts and "combating terrorism overseas" referring to international efforts.[2] Some of these national strategies were specific to combating terrorism, while others involved terrorism to lesser degrees. Table 2 describes the new national strategies related to combating terrorism.

Table 2: National Strategies Related to Combating Terrorism

Strategy	Description of strategy
National Security Strategy of the United States of America • Issued by the President, September 2002	The *National Security* strategy provides a broad framework for strengthening U.S. security in the future. It identifies the national security goals of the United States, describes the foreign policy and military capabilities necessary to achieve those goals, evaluates the current status of these capabilities, and explains how national power will be structured to utilize these capabilities. It devotes a chapter to combating terrorism that focuses on the disruption and destruction of terrorist organizations, the winning of the "war of ideas," the strengthening of homeland security, and the fostering of cooperation with allies and international organizations to combat terrorism.
National Strategy for Homeland Security • Issued by the President, July 2002	The *Homeland Security* strategy addresses the threat of terrorism in the United States by organizing the domestic efforts of federal, state, local, and private organizations. It aligns and focuses homeland security functions into six critical mission areas, set forth as (1) intelligence and warning, (2) border and transportation security, (3) domestic counterterrorism, (4) protecting critical infrastructure and key assets, (5) defending against catastrophic threats, and (6) emergency preparedness and response. Additionally, it describes four foundations that cut across all the mission areas, across all levels of government, and across all sectors of society as being (1) law, (2) science and technology, (3) information sharing and systems, and (4) international cooperation. It also addresses the costs of homeland security and future priorities.
National Strategy for Combating Terrorism • Issued by the President, February 2003	The *Combating Terrorism* strategy elaborates on the terrorism aspects of the *National Security* strategy by expounding on the need to destroy terrorist organizations, win the "war of ideas," and strengthen security at home and abroad. Unlike the *Homeland Security* strategy that focuses on preventing terrorist attacks within the United States, the *Combating Terrorism* strategy focuses on identifying and defusing threats before they reach the borders of the United States. In that sense, although it has defensive elements, this strategy is an offensive strategy to complement the defensive *Homeland Security* strategy.
National Strategy to Combat Weapons of Mass Destruction • Issued by the President, December 2002	The *Weapons of Mass Destruction* strategy presents a national strategy to combat weapons of mass destruction (WMD) through three major efforts: (1) nonproliferation, (2) counterproliferation, and (3) consequence management in WMD incidents. The plan addresses the production and proliferation of WMD among nations, as well as the potential threat of terrorists using WMD agents.

[2] For a more detailed discussion of the definition of terrorism, combating terrorism, and homeland security, see GAO-03-165.

Strategy	Description of strategy
National Strategy for the Physical Protection of Critical Infrastructures and Key Assets • Issued by the President, February 2003	The *Physical Infrastructure* strategy provides a statement of national policy to remain committed to protecting critical infrastructures and key assets from terrorist attacks and is based on eight guiding principles, including establishing responsibility and accountability, encouraging and facilitating partnering among all levels of government and between government and industry, and encouraging market solutions wherever possible and government intervention when needed. The strategy also establishes three strategic objectives. The first is to identify and assure the protection of the most critical assets, systems, and functions, in terms of national level public health and safety, governance, and economic and national security and public confidence. The second is to ensure protection of infrastructures and assets facing specific, imminent threats. The third is to pursue collaborative measures and initiatives to ensure the protection of other potential targets that may become attractive over time.
National Strategy to Secure Cyberspace • Issued by the President, February 2003	The *Secure Cyberspace* strategy is intended to provide an initial framework for both organizing and prioritizing efforts to protect our nation's cyberspace. Also, it is to provide direction to federal departments and agencies that have roles in cyberspace security and to identify steps that state and local governments, private companies and organizations, and individual Americans can take to improve the nation's collective cybersecurity. The strategy is organized according to five national priorities, with major actions and initiatives identified for each. These priorities are: (1) a National Cyberspace Security Response System, (2) a National Cyberspace Security Threat and Vulnerability Reduction Program, (3) a National Cyberspace Security Awareness and Training Program, (4) Securing Governments' Cyberspace, and (5) National Security and International Cyberspace Security Cooperation. In describing the threats to, and vulnerabilities of, our nation's cyberspace, the strategy highlights the potential for damage to U.S. information systems from attacks by terrorist organizations.
2002 National Money Laundering Strategy • Issued by the Secretary of the Treasury and the Attorney General, July 2002	The *Money Laundering* strategy is intended to support planning for the efforts of law enforcement agencies, regulatory officials, the private sector, and overseas entities to combat the laundering of money generated from criminal activities. Although the 2002 strategy still addresses general criminal financial activity, that plan outlines a major governmentwide strategy to combat terrorist financing. The strategy discusses the need to adapt traditional methods of combating money laundering to unconventional tools used by terrorist organizations to finance their operations.

Source: Published national strategies and GAO analysis.

National Strategies Are Broad but Vary in Scope and Detail

These seven national strategies differ from other federal government planning documents, such as agency-specific strategic plans that GPRA requires.[3] These strategies are national in scope, cutting across levels of government and sectors and involving a large number of organizations and entities (i.e., the federal, state, local, and private sectors). In addition, national strategies frequently have international components, and they may be part of a structure of overlapping or supporting national strategies. Furthermore, the federal government does not control many of the sectors, organizations, entities, and resources involved in implementing the national strategies.

[3]P.L. 103-62 (Aug. 3, 1993).

We found that the strategies we studied are organized in a rough hierarchy, with the *National Security* strategy providing an overarching strategy for national security as a whole, including terrorism. The *Homeland Security and Combating Terrorism* strategies provide, respectively, a more specific, defensive approach to combating terrorism at home and an offensive approach to combating terrorism overseas.[4] The other strategies provide further levels of detail on the specific functions related to weapons of mass destruction, cyber security, protection of physical infrastructure, and money laundering. While the national strategies we studied generally overlap in their coverage of terrorism, some contain elements unrelated to terrorism. For example, both the *Secure Cyberspace* and *Money Laundering* strategies include domestic criminal elements that are not necessarily associated with national security or terrorism.[5]

In addition, other executive branch guidance in the form of executive orders or presidential directives elaborates on the national strategies and provides further direction to the implementing parties. Most recently, for instance, the Homeland Security Presidential Directives 7 and 8, issued in December 2003, refine the national strategies with respect to critical infrastructure and national preparedness, respectively. In fact, those presidential directives identify specific priorities and milestones and assign certain responsibilities, which address some of our concerns on the lack of specificity and delineation of clear lines of responsibility in the national strategies. Further down the hierarchy, agency-specific strategic plans and performance plans; federal or agency-level enterprise architectures; and state, local, private and international sector plans provide even further details and guidance to implementing parties. In addition, these plans and reports may address goals and objectives beyond terrorism and homeland security. Figure 1 shows the hierarchy among the national strategies and other plans and guidance.

[4] We recognize that our characterization of these two strategies simplifies a complex relationship. Both strategies contain both defensive and offensive elements. For example, while we characterize the *Homeland Security* strategy as mainly defensive, it includes some offensive initiatives to target and attack terrorist financing, and to track foreign terrorists and bring them to justice. Similarly, while we characterize the *Combating Terrorism* strategy as mainly offensive, it includes some defensive objectives to implement the *Homeland Security* strategy and to protect U.S. citizens abroad.

[5] For example, the *Secure Cyberspace* strategy also covers nonterrorism-related computer hacking, and the *Money Laundering* strategy deals with all types of crimes associated with money laundering, such as drug trafficking.

Figure 1. Hierarchy of National Strategies and Other Plans and Guidance for Combating Terrorism and Homeland Security

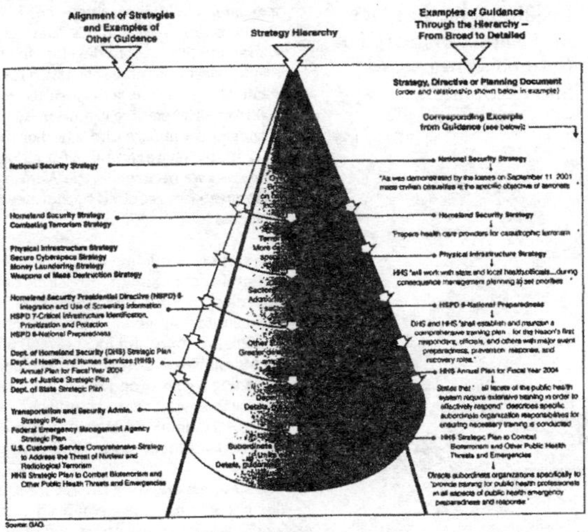

Source: GAO.

GAO Developed A Set of Desirable Characteristics for National Strategies

Because national strategies are not governed by a single, consistent set of requirements, we consulted a variety of public and private sector sources to identify a set of desirable characteristics. Those sources included legislative and executive branch mandates pertaining to the strategies we reviewed, as well as some nonterrorism-related strategies. We also studied GPRA; general literature on strategic planning and performance; and guidance from OMB on the President's Management Agenda. We also gathered published recommendations made by national commissions chartered by Congress; past GAO work; and various research organizations that have commented on national strategies. Based upon this methodology, we identified six characteristics to be desirable for a national strategy, which are described later in this testimony.

No Single Set of Requirements in Place for Characteristics That National Strategies Should Contain

National strategies are not required, either by executive or legislative mandate, to address a single, consistent set of characteristics. Furthermore, we found that there is no commonly accepted set of characteristics used to develop an effective national strategy. Thus to identify desirable characteristics for all national strategies, including those related to terrorism, we consulted numerous sources. First, we identified statutory or executive requirements specific to some of the individual strategies for insight into whether those requirements could be generalized as desirable characteristics for all national strategies. Two of the seven strategies we reviewed—the *National Security* and *Money Laundering* strategies—are required by statutes that mandate specific content elements.[6]

The statute mandating the *Money Laundering* strategy generally calls for the strategy to contain provisions on setting goals, objectives, and priorities; coordinating prevention efforts; specifying detection and prosecution initiatives; and enhancing intergovernmental cooperation (at the federal, state, and local levels) and partnerships between the private sector and law enforcement agencies.[7] In addition, that statute calls for providing 3-year program projections and budget priorities; an assessment of how the budget is to be utilized and its sufficiency; the development of improved communication systems; and evaluations of the effectiveness of policies to combat money laundering and related financial crimes.

The statute mandating the *National Security* strategy calls for the document to provide a comprehensive description and discussion of U.S. worldwide interests, goals, and objectives vital to national security; detail the foreign policy, worldwide commitments, and national defense capabilities necessary to deter aggression and implement the strategy; identify the proposed short- and long-term uses of national power to protect our interests and achieve our goals and objectives; and assess the adequacy of our capabilities to carry out the national strategy.[8]

---

[6] Section 801(b) of the Homeland Security Act of 2002 requires DHS to develop a process for receiving meaningful input from states and localities to assist in the development of a national strategy "for combating terrorism and other homeland security activities," but does not establish specific content elements as do the laws pertaining to the *Money Laundering* and *National Security* strategies.

[7] 31 U.S.C. 5341.

[8] 50 U.S.C. 404a.

However, the requirements set forth in these two statutes, in addition to being different from one another, do not impose any requirements on the five other national strategies we reviewed.

We Developed Characteristics Desirable for National Strategies

Given that there is no established set of requirements for all national strategies—or even the seven related specifically to homeland security and combating terrorism—we developed a set of desirable characteristics by reviewing several sources of information. First, we gathered statutory requirements pertaining to some of the strategies we were asked to assess—namely, the *Money Laundering* and the *National Security* strategies, as mentioned earlier—and legislative and executive branch guidance for other strategies, such as the *National Drug Control Strategy*. We also reviewed GPRA; general literature on strategic planning and performance; and guidance from OMB on the President's Management Agenda. Furthermore, we studied our past reports and testimonies for findings and recommendations pertaining to desirable elements of a national strategy. Similarly, we researched recommendations by national commissions chartered by Congress in recent years on combating terrorism and protecting the homeland—namely, the Bremer, Gilmore, and Hart-Rudman Commissions[9] —and various research organizations that have commented on national strategies.[10] Simultaneously, we consulted widely within GAO to incorporate the most up-to-date thinking on strategic planning; integration across and between government and its partners; implementation; and other related subjects. This included consulting our economists and methodologists to include cost-benefit analysis and other economic criteria. Furthermore, we consulted outside experts from the Bremer and Hart-Rudman Commissions. We used our judgment to develop desirable characteristics based upon their underlying support in legislative or executive guidance and the frequency with which they were cited in other sources. We then grouped similar items together

[9] Even before the terrorist attacks of September 11, 2001, Congress was concerned with the issue of homeland security and had chartered three national commissions, which examined terrorist threats and the government's response to terrorism, and made numerous recommendations. The full names of these commissions are the National Commission on Terrorism (also known as the Bremer Commission), the Advisory Panel to Assess Domestic Response Capabilities for Terrorism Involving Weapons of Mass Destruction (the Gilmore Commission), and the U.S. Commission on National Security/21st Century (the Hart-Rudman Commission).

[10] The research organizations whose work and commentary on homeland security, combating terrorism, and national strategies since 2000 that we primarily reviewed include the ANSER Institute on Homeland Security, RAND Corporation, and Brookings Institution.

in a logical sequence from conception to implementation. This was GAO's first effort to develop desirable characteristics for a national strategy, so they may evolve over time. Table 3 provides a summary of the six characteristics.

Table 3: Summary of Desirable Characteristics for a National Strategy, from Conception to Implementation

Desirable characteristic	Description
Purpose, scope, and methodology	Addresses why the strategy was produced, the scope of its coverage, and the process by which it was developed.
Problem definition and risk assessment	Addresses the particular national problems and threats the strategy is directed towards.
Goals, subordinate objectives, activities, and performance measures	Addresses what the strategy is trying to achieve, steps to achieve those results, as well as the priorities, milestones, and performance measures to gauge results.
Resources, investments, and risk management	Addresses what the strategy will cost, the sources and types of resources and investments needed, and where resources and investments should be targeted based on balancing risk reductions with costs.
Organizational roles, responsibilities, and coordination	Addresses who will be implementing the strategy, what their roles will be compared to others, and mechanisms for them to coordinate their efforts.
Integration and implementation	Addresses how a national strategy relates to other strategies' goals, objectives, and activities, and to subordinate levels of government and their plans to implement the strategy.

Source: GAO data.

We believe a national strategy should ideally contain all of these characteristics. Although the authors of national strategies might organize these characteristics in a variety of ways and/or use different terms, we present them in this order because they flow logically from conception to implementation. Specifically, the strategy's purpose leads to the definition of the problems and risks it intends to address, which in turn leads to specific actions for tackling those problems and risks, allocating and managing the appropriate resources, identifying different organizations' roles and responsibilities, and finally to integrating action among all relevant parties and implementing the strategy.

We describe the desirable characteristics in more detail in the following section, where we evaluate the extent to which the strategies address them. See appendix I for additional details on these characteristics and our scope and methodology in developing them.

National Strategies Address Some, but Not All, of Desirable Characteristics GAO Identified

The seven national strategies related to homeland security and combating terrorism vary considerably in the extent to which they address the desirable characteristics that we identified. All seven strategies we reviewed partially address goals, subordinate objectives, activities, and performance measures. Four of the strategies address problem definition and risk assessment, while one strategy partially addresses that characteristic. And a majority of the strategies at least partially address the four other characteristics: purpose, scope, and methodology; resources, investments, and risk management; organizational roles, responsibilities, and coordination; and integration and implementation. However, none of the strategies addresses all of the elements of resources, investments, and risk management; or integration and implementation. Furthermore, even where the strategies address certain elements of the characteristics, there is room for improvement. For example, while the strategies identify goals, subordinate objectives, and specific activities, they generally do not discuss or identify priorities, milestones, or performance measures—elements that we consider to be desirable for evaluating progress, achieving results, and ensuring effective oversight. On the whole, the *National Strategy for Homeland Security* and the *National Strategy for the Physical Protection of Critical Infrastructure and Key Assets* address the greatest number of the desirable characteristics, while the *National Security Strategy* and the *National Strategy to Combat Weapons of Mass Destruction* address the fewest.

We recognize that strategies themselves are not endpoints, but rather, starting points. In our view, the strengths of some strategies are useful in suggesting ways to enhance the value of other strategies, fill in gaps, speed implementation, guide resource allocations, and provide oversight opportunities. As with any strategic planning effort, implementation is the key. The ultimate measure of these strategies' value will be the extent they are useful as guidance for policy and decision-makers in allocating resources and balancing homeland security priorities with other important, nonhomeland security objectives. It will be important over time to obtain and incorporate feedback from the "user" community as to how the strategies can better provide guidance and how Congress and the administration can identify and remedy impediments to implementation, such as legal, international, jurisdictional, or resource constraints.

Purpose, Scope, and Methodology

This characteristic addresses why the strategy was produced, the scope of its coverage, and the process by which it was developed. For example, a strategy might discuss the specific impetus that led to its being written (or updated), such as statutory requirements, executive mandates, or other events like terrorist attacks. Furthermore, a strategy would enhance clarity by including definitions of key, relevant terms (such as "combating terrorism," and "homeland security" in this context). In addition to describing what it is meant to do and the major functions, mission areas, or activities it covers, a national strategy would ideally address its methodology. For example, a strategy might discuss the principles or theories that guided its development, what organizations or offices drafted the document, whether it was the result of a working group, or which parties were consulted in its development.

Five of the national strategies we evaluated address at least some elements of this characteristic, with four at least partially discussing their overall purpose and scope, and three addressing, to varying degrees, their methodology. For example, the *Homeland Security* strategy explicitly identifies its fundamental objectives, coverage, and how it was developed. It describes itself as a framework to answer four basic questions—such as what is homeland security, and what goals it should pursue—and identifies six "critical mission areas," or homeland security functions, such as intelligence and warning, and border and transportation security. The *Physical Infrastructure*, *Secure Cyberspace*, and *Money Laundering* strategies also use explicit language to define their purposes and scope. For example, the *Physical Infrastructure* strategy identifies its scope as 13 critical sectors (such as agriculture, water, and public health) and five types of key assets (e.g., national monuments and dams). Concerning methodology, the *Homeland Security* strategy explicitly lays out the principles behind its creation and the numerous parties consulted in its development. Similarly, the *Physical Infrastructure* strategy explicitly discusses the guiding principles behind, and the consultations involved in, its creation. The *Combating Terrorism* and *Secure Cyberspace* strategies also describe their guiding principles—and the latter discusses, in even greater detail, the stakeholders involved in its development. And the *Money Laundering* strategy provides its background and highlights changes from the previous version to include terrorist financing.

However, three of the strategies discuss their purpose and scope only in vague terms, and four strategies do not address their methodology at all. For instance, regarding its purpose and scope, the *Weapons of Mass Destruction* strategy says only that, "The United States must pursue a comprehensive strategy to counter the WMD threat in all of its dimensions," without providing any further details. Similarly, while the *National Security* strategy emphasizes the importance of pursuing freedom, peace, and prosperity, it does not state its own purpose or scope. The *Combating Terrorism* strategy also uses vague language, such as "the world must respond and fight this evil," but does not explicitly describe its purpose and scope. In addition, these three strategies, plus the *Money Laundering* strategy, do not discuss who was involved in their development. In our view, a complete description of the purpose, scope, and methodology in a national strategy could make the document more useful to the organizations responsible for implementing the strategy, as well as to oversight organizations, such as the Congress.

Problem Definition and Risk Assessment

This characteristic addresses the particular national problems and threats the strategy is directed towards. Specifically, this means a detailed discussion or definition of the problems the strategy intends to address, their causes, and operating environment. In addition, this characteristic entails a risk assessment, including an analysis of the threats to, and vulnerabilities of, critical assets and operations.[11] If the details of these analyses are classified or preliminary, an unclassified version of the strategy could at least include a broad description of the analyses and stress the importance of risk assessment to implementing parties. A discussion of the quality of data available regarding this characteristic, such as known constraints or deficiencies, would also be useful.

Five of the strategies at least partially address this characteristic. Specifically, five define national problems and the environments in which they occur, while three discuss the importance of assessing risks, threats, and vulnerabilities. For example, the *Combating Terrorism* strategy contains an explicit section on "the nature of the terrorist threat today," which provides some historical background to terrorism, the structure of its leadership, and underlying conditions such as poverty, corruption, religious conflict, and ethnic strife. Similarly, the *Homeland Security, Physical Infrastructure, Secure Cyberspace*, and *Money Laundering* strategies define the problems in their sectors and describe the nature of the terrorist threat. Concerning risk assessment, three of them—the *Homeland Security, Physical Infrastructure*, and *Secure Cyberspace* strategies—stress the importance of national, comprehensive vulnerability

[11] This risk assessment is the first phase of a two-part risk management process. Risk assessment includes a threat assessment, a vulnerability assessment, and a criticality assessment. For a more in-depth discussion of these subjects, see U.S. General Accounting Office, *Homeland Security: Key Elements of a Risk Management Approach*, GAO-02-150T (Washington, D.C.: Oct.12, 2002). The second aspect of risk management is discussed below in the "Resources, Investments and Risk Management" characteristic. It consists of taking the information from the risk assessment and making management decisions about resource allocations to minimize risks and maximize returns on resources expended.

assessments of all critical infrastructures and key assets, setting the stage for risk management. The *Homeland Security* strategy contains an explicit "threat and vulnerability" section that provides many details, such as defining the different ways and means for terrorist attacks. This strategy also stresses the importance of comprehensive vulnerability assessments of all critical infrastructures and key assets, saying they "are important from a planning perspective in that they enable authorities to evaluate the potential effects of an attack on a given facility or sector, and then to invest accordingly in protecting such facilities and sectors."

However, two strategies do not address this characteristic. The *National Security* strategy says the war against terrorism is global and that "The enemy is not a single political regime or person or religion or ideology," but provides no further definition of the problems it seeks to address. Similarly, the *Weapons of Mass Destruction* strategy states that such weapons represent a great security challenge when in the possession of hostile states and terrorists, and that some terrorism-supporting states already possess such weapons, but provides no details defining the threat. Furthermore, while some of the strategies say that intelligence gathering must be strengthened, the strategies generally do not address limitations in collecting data. That is, few of the strategies discuss the difficulties of collecting intelligence on terrorist organizations, plans, and tactics. In our view, more specific information on both problem definition and risk assessment in many of the strategies would give the responsible parties better guidance to implement those strategies. For example, we recently recommended that future *Money Laundering* strategies link to periodic assessments of threats and risks, which would provide a basis for ensuring that clear priorities are established and focused on the areas of greatest need.[12]

Without necessarily prescribing in detail the "solution," better problem definition and risk assessment also provide greater latitude to responsible parties to develop innovative approaches that are tailored to the needs of specific regions or sectors—and are able to be implemented as a practical matter, given fiscal, human capital, and other limitations. For example, better problem definition or risk assessment can foster regional approaches or cooperative agreements, and stimulate the development of national systems or management standards to link the capabilities of the

[12] See U.S. General Accounting Office, *Combating Money Laundering: Opportunities Exist to Improve the National Strategy*, GAO-03-813 (Washington, D.C.: Sept. 2003).

responsible parties in a more effective manner. Such assessments help identify desired goals and "end-states" without "one-size-fits-all" solutions.

Goals, Subordinate Objectives, Activities, and Performance Measures

This characteristic addresses what the national strategy strives to achieve and the steps needed to garner those results, as well as the priorities, milestones, and performance measures to gauge results. At the highest level, this could be a description of an ideal "end-state," followed by a logical hierarchy of major goals, subordinate objectives, and specific activities to achieve results. In addition, it would be helpful if the strategy discussed the importance of implementing parties' establishing priorities, milestones, and performance measures to help ensure accountability. Ideally, a national strategy would set clear desired results and priorities, specific milestones, and outcome-related performance measures while giving implementing parties flexibility to pursue and achieve those results within a reasonable timeframe. If significant limitations on performance measures exist, other parts of the strategy might address plans to obtain better data or measurements, such as national standards or indicators of preparedness. For example, national strategies related to terrorism might discuss the lack of national indicators or standards for emergency preparedness against attacks.

All seven national strategies partially address this characteristic by identifying their individual, high-level goals, subordinate objectives, and specific activities to achieve results.[13] For example, the *Homeland Security* strategy identifies three major goals—prevent terrorist attacks, reduce vulnerability, and minimize damage and recover from attacks—which are underpinned by six objectives (called critical mission areas), such as intelligence and warning, and border and transportation security. Those objectives in turn, have anywhere from 5 to 12 accompanying activities apiece. Figure 2 illustrates an example of an overall goal, subordinate objective, and specific activity in the *Homeland Security* strategy.

[13] The strategies differ in their terminology for goals, objectives, and activities. For example, some strategies refer to their top-level vision as "goals," while others describe that as "objectives." The same is true at the next level of support—some are called objectives, while others are "priorities" or "critical mission areas"—and at the most detailed level of activities (alternatively called "priorities" or "initiatives"). For the purpose of consistency in this testimony, we are using the terms "goals," "subordinate objectives," and "activities" (in order of broad to specific).

Figure 2: The Homeland Security strategy contains an overall goal on recovering from terrorist attacks, a subordinate objective on emergency preparedness and response, and a specific initiative to prepare for chemical, biological, and nuclear decontamination

Source: GAO

Similarly, the *Combating Terrorism* strategy contains four overarching goals: defeat terrorists and their organizations; deny sponsorship, support, and sanctuary to terrorists; diminish the underlying conditions that terrorists seek to exploit; and defend U.S. citizens and interests at home and abroad. These goals are broken down into 15 objectives, such as identifying terrorists and terrorist organizations, and are further supported by one to four activities each. Concerning milestones, the *Money Laundering* strategy provides a few deadlines for specific activities, such as the Departments of Treasury and Justice conducting a study by April 2003 on how the Internet could be used by terrorist groups to raise money. In addition, the *Homeland Security* strategy calls for DHS to develop and coordinate implementation of a comprehensive national plan to protect infrastructure against terrorist attacks, building on baseline protection plans due by the end of fiscal year 2002.[14] Regarding performance measures, the *Homeland Security* and *Money Laundering* strategies provide some general language on the subject. For example, the former says that, "Every department or agency will create benchmarks and other performance measures by which we can evaluate our progress and allocate future resources." And the latter says that methods for measuring performance should be consistent with the President's Management Agenda, and that the Department of the Treasury will develop a "traffic light" scorecard to track performance and assess how well the strategies' initiatives are being implemented.

However, the strategies do not address this characteristic in that they generally lack priorities, milestones, or performance measures. Regarding priorities, only the *Homeland Security* strategy identifies a priority order by stressing the importance of four specific activities in the fiscal year 2003 budget. Five strategies do not designate specific priorities; and the *Money Laundering* strategy, as highlighted in our recent report, identifies more priorities than can be achieved in a reasonable timeframe and does not rank them in order of importance.[15] Concerning performance measures, only two of them—the *Homeland Security* and *Money Laundering* strategies—explicitly stress the importance of measuring performance or identify specific measures. As we said in an earlier testimony, the *Homeland Security* strategy's initiatives often do not provide a baseline set of performance goals and measures upon which to

[14] The Homeland Security Act of 2002 requires DHS to develop a comprehensive national plan for securing the key resources and critical infrastructure of the United States (P.L. 107-296, sec. 201(d)(5). Consistent with the Act, section (27) of the Homeland Security Presidential Directive 7 requires the Secretary of Homeland Security to complete a comprehensive, integrated National Plan for Critical Infrastructure and Key Resources Protection that outlines national goals, objectives, milestones, and key initiatives by December 2004.

[15] See GAO-03-813.

assess and improve preparedness.[16] Similarly, we recently recommended that future *Money Laundering* strategies require the principal agencies to develop outcome-related performance measures that are linked to goals and objectives.[17] Also, we previously reported that neither the *Physical Infrastructure* nor the *Secure Cyberspace* strategies indicate timeframes or milestones for their overall implementation or for accomplishing specific actions or initiatives; nor do they establish performance measures for which entities can be held responsible.[18] We believe a better identification of priorities, milestones, and performance measures would aid implementing parties in achieving results in specific timeframes—and would enable more effective oversight and accountability.

Resources, Investments, and Risk Management

This characteristic addresses what the strategy will cost, the sources and types of resources and investments associated with the strategy, and where those resources and investments should be targeted. Ideally, a strategy would also identify criteria and appropriate mechanisms to allocate resources, such as grants, in-kind services, loans, and user fees, based on identified needs. Alternatively, the strategy might identify appropriate "tools of government," such as regulations, tax incentives, and standards, to mandate or stimulate nonfederal organizations to use their unique resources. Furthermore, a national strategy would ideally elaborate on the risk assessment mentioned earlier and give guidance to implementing parties to manage their resources and investments accordingly—and begin to address the difficult but critical issues about who pays, and how such efforts will be funded and sustained in the future.

Four of the strategies we evaluated partially address this characteristic by identifying numerous resource and investment needs to achieve their goals and objectives, and by discussing, to varying degrees, risk management. The *Homeland Security* strategy goes even farther, devoting a chapter to this topic in which it identifies a general principle to allocate homeland security investments based upon balancing risk reductions and costs. For example, the strategy states, "Decisions on homeland security activities

[16] See U.S. General Accounting Office, *Homeland Security: Effective Intergovernmental Coordination is Key to Success* GAO-02- 1011T (Washington, D.C.: August 2002).

[17] See GAO-03-813.

[18] See U.S. General Accounting Office, *Homeland Security: Information Sharing Responsibilities, Challenges, and Key Management Issues*, GAO-03-1165T (Washington, D.C.: September 2003).

and spending must achieve two overarching goals: to devote the right amount of scarce resources to homeland security and to spend these resources on the right activities." In addition, the *Homeland Security* strategy cites the concept that "the federal government will provide an incentive to minimize costs and reward innovation by permitting maximum flexibility in meeting those objectives." While the *Homeland Security* strategy cites these principles, it still provides relatively few details on the types and levels of resources associated with implementation. The *Physical Infrastructure* strategy also partially addresses this characteristic by identifying planning and resource allocation as one of its five objectives—and by stressing the importance of incentives for private organizations, and market solutions where appropriate. And the *Secure Cyberspace* strategy is one of only two strategies (the other being the *Homeland Security* strategy) to link some of its investment requests—such as completing the installation of the Cyber Warning and Information Network in key government operation centers—to the fiscal 2003 budget. The *Money Laundering* strategy also briefly discusses the importance of cost-benefit analysis of asset forfeiture strategies "so that future programs can allocate resources where they are most needed and productive." Figure 3 shows spending for combating terrorism by federal agency.

Figure 3: Budget Authority for Combating Terrorism by Agency for Fiscal Year 2004 (total budget authority is $52,732 million)

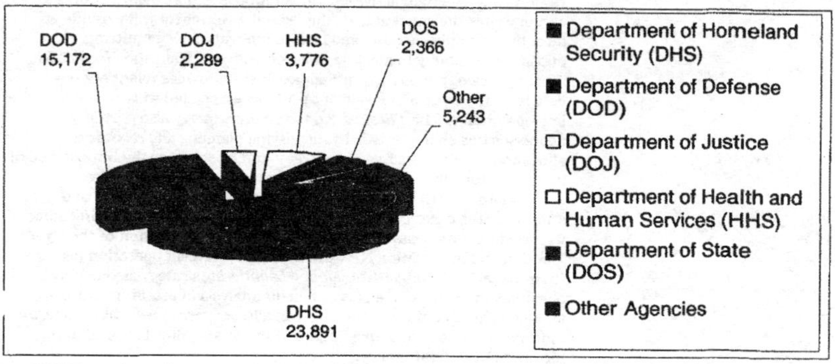

Source: OMB 2003 Report on Combating Terrorism.

Note: "Other Agencies" includes the Departments of Energy ($1,588 million), Agriculture ($368 million), Transportation ($283 million), Commerce ($153 million), Veterans Affairs ($145 million), Interior ($115 million), Treasury ($90 million), Labor ($67 million), Housing and Urban Development ($2 million), and 18 other independent agencies (totaling $2,432 million).

Regarding risk management, the *Homeland Security* strategy makes explicit reference to the subject, such as when it says, "The national effort to enhance homeland security will yield tremendous benefits and entail substantial financial and other costs." The *Physical Infrastructure* and *Secure Cyberspace* strategies also mention risk management, building on their aforementioned sections on risk assessment. In the former, for instance, increased sharing of risk-management expertise between the public and private sectors is an activity identified under the planning and resource allocation objective.

On the other hand, three of the strategies—the *National Security*, *Combating Terrorism*, and *Weapons of Mass Destruction* strategies—do not explicitly address either resource and investment needs or risk management. And of those that partially address this characteristic, only two—the *Homeland Security* and *Physical Infrastructure* strategies—provide explicit guidance or principles concerning resource allocation. Along those lines, none of the strategies provides cost estimates for implementation in the aggregate, nor for specific goals, objectives, or activities. In addition, none of the strategies contains distinct chapters or sections, or detailed discussions of risk management. In our view, more guidance on resource, investment, and risk management would help implementing parties allocate resources and investments according to priorities and constraints, track costs and performance, and shift such investments and resources as appropriate. Such guidance would also assist Congress and the administration in developing more effective federal programs to stimulate desired investments, enhance preparedness, and leverage finite resources.

Organizational Roles, Responsibilities, and Coordination

This characteristic addresses which organizations will implement the strategy, their roles and responsibilities, and mechanisms for coordinating their efforts. It helps answer the fundamental question about who is in charge, not only during times of crisis, but also during all phases of homeland security and combating terrorism efforts: prevention, vulnerability reduction, and response and recovery. This characteristic entails identifying the specific federal departments, agencies, or offices involved and, where appropriate, the different sectors, such as state, local, private, or international sectors. A strategy would ideally clarify implementing organizations' relationships in terms of leading, supporting, and partnering.[19] In addition, a strategy could describe the organizations that will provide the overall framework for accountability and oversight, such as the National Security Council, Homeland Security Council, OMB, Congress, or other organizations. Furthermore, a strategy might also identify specific processes for coordination and collaboration between sectors and organizations—and address how any conflicts would be resolved. For example, a strategy might also provide for some mechanism to ensure that the parties are prepared to fulfill their assigned responsibilities and use their available resources appropriately to enhance their capabilities and preparedness.

Six strategies at least partially address this characteristic. Specifically, two of them—the *Homeland Security* and *Physical Infrastructure* strategies—contain distinct chapters on "organizing," which discuss roles and responsibilities among the federal, state, local, private, and international

[19] By "partnering," we refer to shared, or joint, responsibilities between implementing parties where there is otherwise no clear or established hierarchy of lead and support functions.

sectors.[20] Furthermore, those two strategies, plus the *Secure Cyberspace* and *Money Laundering* strategies, frequently designate lead, and sometimes support, roles by objective, sector, or even specific activity.[21] Regarding accountability and oversight, the *Combating Terrorism* strategy identifies the creation of an international standard as one of its objectives, and the *Homeland Security* and *Physical Infrastructure* strategies highlight the importance of accountability. And concerning coordination between implementing parties, the *Homeland Security* and *Money Laundering* strategies designate some specific tools or processes (e.g., steering committee or task force), and the *Physical Infrastructure* strategy identifies the need to create collaborative mechanisms for government-industry planning; it also designates DHS as the primary liaison and facilitator for cooperation between all relevant parties.

On the other hand, the *National Security* strategy does not address this characteristic at all, and there is room for improvement in the other six strategies as well. For example, many of the references to U.S. roles and responsibilities in the *National Security* and *Combating Terrorism* strategies simply designate "the United States," rather than a specific federal agency, level of government, or sector. Thus those two strategies do not identify lead, support, and partner roles like the other strategies do. In addition, none of the strategies defines an overarching accountability or oversight framework, and five of the strategies do not identify specific tools or processes for coordination. For example, we recently recommended that future *Money Laundering* strategies address, among other things, strengthening the leadership structure and establishing a mechanism to resolve disputes among agencies and ensure accountability for implementation.[22] Also, we previously reported that neither the *Physical Infrastructure* nor the *Secure Cyberspace* strategies adequately define the roles, responsibilities, and relationships among the key critical infrastructure protection organizations, including state and local governments and the private sector.[23] The inclusion of these subjects in a

[20] The *Homeland Security* strategy places many responsibilities on DHS, which had not been created yet when the strategy was published.

[21] The unclassified *Weapons of Mass Destruction* strategy outlines only a few specific responsibilities for the Homeland Security Council, National Security Council, and Department of State. However, its classified version contains more relevant details, which cannot be addressed in this unclassified statement.

[22] See GAO-03-813.

[23] See GAO-03-1165T.

national strategy would be useful to agencies and other stakeholders in fostering coordination and clarifying specific roles, particularly where there is overlap, and thus enhancing both implementation and accountability.

Integration and Implementation

This characteristic addresses both how a national strategy relates to other strategies' goals, objectives, and activities—and to subordinate levels of government and their plans to implement the strategy. For example, a national strategy could discuss how its scope complements, expands upon, or overlaps with other national strategies, such as transportation infrastructure recapitalization or energy reliability. Similarly, related strategies could highlight their common or shared goals, subordinate objectives, and activities. In addition, a national strategy could address its relationship with relevant documents from implementing organizations, such as the strategic plans, annual performance plans, or annual performance reports required of federal agencies by GPRA. A strategy might also discuss, as appropriate, various strategies and plans produced by the state, local, private, or international sectors. It could also provide guidance such as the development of national standards to link together more effectively the roles, responsibilities, and capabilities of the implementing parties.

Five of the strategies address certain elements of this characteristic. Specifically, in terms of integration, the *Homeland Security* strategy states that it complements the *National Security* strategy in providing a framework for other security-related strategies and, in this vein, lays out goals, objectives, and mission areas that are shared with other strategies. The *Combating Terrorism*, *Weapons of Mass Destruction*, and *Secure Cyberspace* strategies also address integration by discussing the importance of other strategies and their complementary relationships. The *Homeland Security* and *Physical Infrastructure* strategies also provide some language on this subject, such as the latter's statement that DHS will collaborate with state and local governments as well as other federal agencies and the private sector to implement structures and processes for protecting assets and infrastructure. Regarding implementation, the *Homeland Security* strategy contains a distinct section on the subject, acknowledging that executive branch agencies need to issue detailed plans for the strategy's initiatives. And the *Money Laundering* strategy, for many of its activities, lists specific "action items" for agencies to implement. Two other strategies—the *Physical Infrastructure* and *Secure Cyberspace* strategies—make some general references to implementation. For example, the former says that "DHS and designated federal lead departments and agencies will prepare detailed implementation plans to support the activities outlined."

However, one of the strategies we reviewed—the *National Security* strategy—does not address this characteristic. It does not define its relationship to the other strategies; nor does it (along with the *Combating Terrorism, Weapons of Mass Destruction, Secure Cyberspace*, and *Money Laundering* strategies) address their relationship with other plans by federal, state, local, and other implementing parties. Furthermore, three strategies—the *National Security, Combating Terrorism*, and *Weapons of Mass Destruction* strategies—do not explicitly address implementation, and none of the strategies provides detailed guidance on the subject. We believe more information on this characteristic in a national strategy would build on the aforementioned organizational roles and responsibilities—and thus further clarify the relationships between various implementing parties, both vertically and horizontally. This, in turn, would foster effective implementation and accountability.

Concluding Observations

The seven national strategies addressing homeland security and combating terrorism that we discuss in this testimony were developed to help the United States respond to an array of potential threats brought sharply into focus after the terrorist attacks of September 11, 2001. We recognize that these strategies were issued to meet a variety of homeland security needs and, furthermore, that they were not required, for the most part, to address the characteristics that we consider to be desirable. In addition, we do not expect all of the strategies to provide the same degree of detail because of their different scopes; for example, we consider it appropriate for the *National Security* strategy to contain fewer specifics than the *Physical Infrastructure* or *Money Laundering* strategies. Nonetheless, in our view, it would be useful for all of the strategies to address each of the characteristics, which logically flow from conception to implementation, in order to provide guidance to the federal agencies and other parties responsible for achieving results, evaluating progress, and ensuring accountability. Even where the strategies address our characteristics, we have identified potential areas for improvement. The numerous examples that I have cited today of the characteristics' inclusion in the national strategies may serve as a model for future versions of these and other strategies.

The ultimate value of these strategies will be determined through time as the strategies are implemented by the federal, state, local, private, and international sectors—and as homeland security actions are embedded or integrated into ongoing governmental and private sector missions in sustainable and balanced ways. To achieve these goals, it will continue to be important to solicit the feedback and input from all responsible parties—legislative, federal, state, local, private, and international—and to incorporate this information to better achieve the parties' shared goals of improved homeland security and national preparedness. We will continue our work for the Subcommittee to evaluate these national strategies and their implementation. In the coming weeks, we look forward to reporting on (1) the extent that these strategies address recommendations by national commissions and GAO, (2) the extent to which implementing agencies are incorporating the national strategies into their own plans, and (3) the challenges faced in implementing these national strategies.

Mr. Chairman, this concludes my prepared statement. I will be pleased to respond to any questions that you or other members of the Subcommittee may have.

GAO Contact and Staff Acknowledgments

GAO Contact

Randall Yim at (202) 512-6787

GAO Acknowledgments

Individuals making key contributions to this statement include Stephen L. Caldwell, Sharon Caudle, Josey Ballenger, Heather MacLeod, Jared Hermalin, Wayne A. Ekblad, Amy Bernstein, and Christine Davis.

Appendix I: Scope and Methodology

This appendix describes how we developed the characteristics that we consider to be desirable for a national strategy and how we used them to evaluate the national strategies related to combating terrorism and homeland security.

Developing Desirable Characteristics for a National Strategy

There are no legislative or executive mandates identifying a uniform set of required or desirable characteristics for all national strategies, including those related to combating terrorism and homeland security. While two of the seven strategies we reviewed—the *National Security* and *Money Laundering* strategies—are required by statutes to include specific content elements, the requirements set forth in these two statutes, in addition to being different from one another, do not levy any requirements on the five other national strategies we reviewed.

Given that there is no established set of requirements for all national strategies—or even the seven related specifically to combating terrorism and homeland security—we identified a set of desirable characteristics by reviewing several sources of information. First, we gathered statutory requirements pertaining to some of the strategies we were asked to assess—namely, the *Money Laundering* and *National Security* strategies, as mentioned earlier—as well as legislative and executive branch guidance for other strategies, such as the *National Drug Control Strategy*. We also consulted the Government Performance and Results Act (GPRA) of 1993; general literature on strategic planning and performance;[1] and guidance from the Office of Management and Budget (OMB) on the President's Management Agenda. In addition, we studied our past reports and testimonies for findings and recommendations pertaining to desirable elements of a national strategy. Similarly, we researched recommendations by national commissions chartered by Congress in recent years on combating terrorism and protecting the homeland—namely, the Bremer, Gilmore, and Hart-Rudman Commissions—and various research organizations that have commented on national strategies, such as the ANSER Institute on Homeland Security, RAND Corporation, and Brookings Institution.

[1] Examples of such literature include John M. Bryson's book *Strategic Planning for Public and Nonprofit Organizations: A Guide to Strengthening and Sustaining Organizational Achievement* (Jossey-Bass, 1995) and Edward Filiberti's article, *National Strategic Guidance: Do We Need a Standard Format?* (Parameters, U.S. Army War College, Autumn 1995).

Simultaneously, we consulted widely within GAO to incorporate the most up-to-date thinking on strategic planning, integration across and between government and its partners, implementation, and other related subjects. This included consulting our economists and methodologists to include cost-benefit analysis and other economic factors. Furthermore, we consulted outside experts from the Bremer and Hart-Rudman Commissions.

We used our judgment to develop desirable characteristics based on their underlying support in legislative or executive guidance and the frequency with which they were cited in other sources. We then grouped similar items together in a logical sequence, from conception to implementation. This is our first effort to develop desirable characteristics for an effective national strategy, so they may evolve over time. The desirable characteristics are:

- Purpose, scope, and methodology.
- Problem definition and risk assessment.
- Goals, subordinate objectives, activities, and performance measures.
- Resources, investments, and risk management.
- Organizational roles, responsibilities, and coordination.
- Integration and implementation.

Later in this appendix, we provide a more detailed description of the six characteristics, plus examples of elements that a strategy might include to address them. We believe a national strategy should ideally contain all of these characteristics. Although the authors of national strategies might organize them in a variety of ways and/or use different terms, we present the characteristics in this order as a logical flow from conception to implementation. Specifically, the strategy's purpose leads to the definition of the problems and risks it intends to address, which in turn leads to specific actions for tackling those problems and risks, allocating and managing the appropriate resources, identifying different organizations' roles responsibilities and, finally, to integrating action among all relevant parties and implementing the strategy.

One challenge we encountered in identifying and applying these characteristics was determining the appropriate level of specificity a national strategy might contain. We found that there was no consensus on this issue among the sources and experts we consulted. Furthermore, the strategies we reviewed vary in their scope of coverage—some are broad strategies, while others focus on implementation—and thus their level of detail varies.[2] We recognize that by their nature, national strategies are intended to provide broad direction and guidance—rather than be prescriptive, detailed mandates—to the relevant implementing parties. Thus it is unrealistic to expect all of the national strategies to provide details on each and every key characteristic we identified. Nonetheless, we believe the more detail a strategy provides, the easier it is for the responsible parties to implement it and achieve its goals. Table 4 provides the desirable characteristics and examples of their elements.

Table 4: GAO Desirable Characteristics for a National Strategy

Desirable Characteristic	Brief description	Examples of elements
Purpose, scope, and methodology	Addresses why the strategy was produced, the scope of its coverage, and the process by which it was developed.	• Statement of broad or narrow purpose, as appropriate. • How it compares and contrasts with other national strategies. • What major functions, mission areas, or activities it covers. • Principles or theories that guided its development. • Impetus for strategy, e.g. statutory requirement or event. • Process to produce strategy, e.g. interagency task force; state, local, or private input. • Definition of key terms.
Problem definition and risk assessment	Addresses the particular national problems and threats the strategy is directed towards.	• Discussion or definition of problems, their causes, and operating environment. • Risk assessment, including an analysis of threats and vulnerabilities. • Quality of data available, e.g. constraints, deficiencies, and "unknowns."
Goals, subordinate objectives, activities, and performance measures	Addresses what the strategy is trying to achieve, steps to achieve those results, as well as the priorities, milestones, and performance measures to gauge results.	• Overall results desired, i.e. "end-state." • Hierarchy of strategic goals and subordinate objectives. • Specific activities to achieve results. • Priorities, milestones, and outcome-related performance measures. • Specific performance measures. • Process for monitoring and reporting on progress. • Limitations on progress indicators.

[2] For example, the strategies range from the high-level, "grand" strategy (e.g., the *National Security* strategy) to the mid-level strategies specific to terrorism (e.g., the *Homeland Security* and *Combating Terrorism* strategies) and, finally, to the more detailed, sector- or function-specific strategies geared towards implementation (e.g., the *Secure Cyberspace*, and *Money Laundering* strategies).

Desirable Characteristic	Brief description	Examples of elements
Resources, investments, and risk management	Addresses what the strategy will cost, the sources and types of resources and investments needed, and where resources and investments should be targeted by balancing risk reductions and costs.	• Resources and investments associated with the strategy. • Types of resources required, such as budgetary, human capital, information technology, research and development, contracts. • Sources of resources, e.g., federal, state, local, and private. • Economic principles, such as balancing benefits and costs. • Resource allocation mechanisms, such as grants, in-kind services, loans, or user fees. • "Tools of government," e.g., mandates or incentives to spur action. • Importance of fiscal discipline. • Linkage to other resource documents, e.g. federal budget. • Risk management principles.
Organizational roles, responsibilities, and coordination	Addresses who will be implementing the strategy, what their roles will be compared to others, and mechanisms for them to coordinate their efforts.	• Roles and responsibilities of specific federal agencies, departments, or offices. • Roles and responsibilities of state, local, private, and international sectors. • Lead, support, and partner roles and responsibilities. • Accountability and oversight framework. • Potential changes to current organizational structure. • Specific processes for coordination and collaboration. • How conflicts will be resolved.
Integration and implementation	Addresses how a national strategy relates to other strategies' goals, objectives and activities – and to subordinate levels of government and their plans to implement the strategy.	• Integration with other national strategies (horizontal). • Integration with relevant documents from implementing organizations (vertical). • Details on specific federal, state, local, or private strategies and plans. • Implementation guidance. • Details on subordinate strategies and plans for implementation, e.g., human capital, and enterprise architecture.

Source: GAO.

The following sections provide more detail on the six characteristics and our support of each of them.

Purpose, Scope, and Methodology

This characteristic addresses why the strategy was produced, the scope of its coverage, and the process by which it was developed. For example, a strategy might discuss the specific impetus that led to its being written (or updated), such as statutory requirements, executive mandates, or other events like terrorist attacks. Furthermore, a strategy would enhance clarity by including definitions of key, relevant terms (such as "homeland security" and "combating terrorism," in this context). In addition to describing what it is meant to do and the major functions, mission areas, or activities it covers, a national strategy would ideally address its methodology. For example, a strategy might discuss the principles or theories that guided its development, what organizations or offices drafted the document, whether it was the result of a working group, or which parties were consulted in its development.

We found support for this characteristic in legislation mandating two of the seven national strategies as well as by related legislation, executive orders, and GAO and policy research organization publications. For example, provisions relating to "purpose, scope, and methodology" appear in the statutes mandating the *National Security*[3] and *Money Laundering* strategies[4] (e.g., the statute requiring the *Money Laundering* strategy sets forth 12 areas that the strategy shall address.) Other legislative and executive branch guidance justifying the inclusion of this characteristic in our typology include: statutory requirements and related government publications describing the required purpose, scope, and methodology for the *National Drug Control Strategy*;[5] GPRA legislation calling for a comprehensive mission statement in agency strategic plans;[6] and an executive order determining the purpose and scope of a national council/strategy on information infrastructure.[7] In addition, at least two of our testimonies have directly addressed the relevant purpose and scope issues to be included within a homeland security strategy (e.g., the strategy is to be "national" in scope; its purpose is to include setting overall priorities and goals for homeland security). [8] But, we also pointed out in a 2002 testimony, that based upon interviews with officials at a dozen federal agencies, a broadly accepted definition of homeland security does not exist and that further clarification is needed.[9] The Gilmore Commission and ANSER Institute for Homeland Security have also

[3] 50 U.S.C. 404a.

[4] 31 U.S.C. 5341.

[5] See Section 1005 of the Anti-Drug Abuse Act of 1988, P.L. 100-690 (Nov. 18, 1988).

[6] See P.L. 103-62, sec. 3 (Aug. 3, 1993).

[7] Executive Order 12864 (Sept. 15, 1993).

[8] See U.S. General Accounting Office, *Homeland Security: Key Elements to Unify Efforts Are Underway but Uncertainty Remains*, GAO-02-610 (Washington, D.C.: June, 2002), p. 9; and *Homeland Security: Proposal for Cabinet Agency Has Merit, But Implementation Will be Pivotal to Success*, GAO-02-886T (Washington, D.C.: June 25, 2002), p. 4.

[9] See U.S. General Accounting Office, *Homeland Security: Progress Made; More Direction and Partnership Sought*, GAO-02-490T (Washington, D.C.: Mar. 12, 2002), p. 9.

addressed aspects of "purpose, scope, and methodology" issues that need to be addressed in a national strategy (e.g., the Gilmore Commission indicates that the strategy should be functionally comprehensive and address the full spectrum of the nation's efforts against terrorism).[10]

Problem Definition and Risk Assessment

This characteristic addresses the particular national problems and threats the strategy is directed towards. Specifically, this means a detailed discussion or definition of the problems the strategy intends to address, their causes, and operating environment. In addition, this characteristic entails a risk assessment, including an analysis of the threats to, and vulnerabilities of, critical assets and operations.[11] If the details of these analyses are classified or preliminary, an unclassified version of the strategy could at least include a broad description of the analyses and stress the importance of risk assessment to implementing parties. A discussion of the quality of data available regarding this characteristic, such as known constraints or deficiencies, would also be useful.

Again, we found support for this characteristic in a variety of sources. While we have not identified any legislation that requires use of this characteristic in the national strategies on combating terrorism and homeland security that we reviewed, the importance of this characteristic is supported by the Homeland Security Act of 2002, as well as other legislation, presidential directives, and GAO and policy research organization publications. For example, the Homeland Security Act of 2002 directs the Department of Homeland Security (DHS) to conduct comprehensive assessments of vulnerabilities, including risk assessments;[12] GPRA requires the identification of key factors external to

[10] Second Annual Report to The President and The Congress Of the Advisory Panel to Assess Domestic Response Capabilities for Terrorism Involving Weapons of Mass Destruction (aka Gilmore Commission), *II. Toward A National Strategy For Combating Terrorism* (Dec. 15, 2000), p. 4; Ruth David, *Homeland Security: Building A National Strategy*, The Bridge, 32, 1 (Spring, 2002), p. 2.

[11] This risk assessment is the first phase of a two-part risk management process. Risk assessment includes a threat assessment, a vulnerability assessment, and a "criticality" analysis. For a more in-depth discussion of these subjects, see *Homeland Security: Key Elements of a Risk Management Approach*, GAO-02-150T (Washington, D.C.: Oct.12, 2002). The second aspect of risk management is discussed in the "Resources, Investments and Risk Management" characteristics. It consists of taking the information from the risk assessment and making management decisions about resource allocations to minimize risks and maximize returns on resources expended.

[12] P.L. 107-296, sec. 201(d)(2).

an agency that can significantly impact that agency's attainment of its goals and objectives; [13] Homeland Security Presidential Directive (HSPD) 7, which addresses critical infrastructure protection, contains a background section that defines problem areas, and assesses the national risk potential if such problem areas are not effectively addressed. Likewise, an earlier critical infrastructure directive, Presidential Decision Directive (PDD) 63 defines the growing concern about the nation's vulnerability.[14] Additionally, we testified in 2002 that use of common definitions promotes more effective intergovernmental operations and more accurate monitoring of expenditures, thereby eliminating problematic concerns.[15] We also said that a national homeland security strategy should be based on a comprehensive national threat and risk assessment.[16] The Gilmore Commission, ANSER, and RAND have all suggested the need to conduct threat assessments to the homeland.[17]

Goals, Subordinate Objectives, Activities, and Performance Measures

This characteristic addresses what the national strategy strives to achieve and the steps needed to garner those results, as well as the priorities, milestones, and performance measures to gauge results. At the highest level, this could be a description of an ideal "end-state," followed by a logical hierarchy of major goals, subordinate objectives, and specific activities to achieve results. In addition, it would be helpful if the strategy discussed the importance of implementing parties' efforts to establish priorities, milestones, and performance measures which help ensure accountability. Ideally, a national strategy would set clear desired results and priorities, specific milestones, and outcome-related performance measures while giving implementing parties flexibility to pursue and

[13] P.L. 103-62, sec. 3.

[14] *See* Homeland Security Presidential Directive/HSPD-7, Critical Infrastructure Identification, Prioritization and Protection, Dec. 17, 2003, and Presidential Decision Direction/NSC-63, Critical Infrastructure Protection, May 22, 1998. HSPD-7 states that it supersedes PDD/NS C-63 to the extent of any inconsistency.

[15] See GAO-02-490T.

[16] See U.S. General Accounting Office, *Homeland Security: A Framework for Addressing the Nation's Efforts*, GAO-01-1158T (Washington, D.C.: September 21, 2001), p. 1.

[17] First Annual Report to The President and The Congress Of the Advisory Panel To Assess Domestic Response Capabilities For Terrorism Involving Weapons Of Mass Destruction (aka Gilmore Commission), *I. Assessing the Threat* (December 15, 1999), p. 55; Ruth David, *Homeland Security: Building a National Strategy*, The Bridge, 32, 1 (Spring, 2002), p. 4; Bruce Hoffman, *Combating Terrorism: In Search of a National Strategy* RAND Corporation, CT-175, March 2001, pp. 3,6-7.

achieve those results within a reasonable timeframe. If significant limitations on performance measures exist, other parts of the strategy might address plans to obtain better data or measurements, such as national standards or indicators of preparedness.[18] For example, national strategies related to terrorism might discuss the lack of national indicators or standards for emergency preparedness against attacks.

As in the case of the first characteristic, we found support for this characteristic in legislation mandating the *Money Laundering* and *National Security* strategies, as well as support derived from related legislation, presidential directive, the President's Management Agenda, and GAO and policy research organization publications. Both the *National Security* strategy and the *Money Laundering* strategy statutes emphasize the need for goals and objectives, as well as operational initiatives to promote those goals and objectives. There is also related legislative and executive supporting guidance for this characteristic in the following: the *National Drug Control Strategy* legislation, which requires a complete list of goals, objectives, and priorities;[19] the Homeland Security Act of 2002, which requires DHS to develop, in connection with a national terrorism countermeasures strategy, comprehensive, research-based definable goals and annual measurable objectives and specific targets to accomplish and evaluate such goals;[20] GPRA, which requires federal agencies to set goals and objectives in their strategic plans;[21] PDD 63, which includes a statement of presidential intent and national goals; [22] and the President's Management Agenda of FY2002,[23] which describes OMB's work regarding program objectives. Additionally, we testified that a national strategy should establish goals, objectives, and performance measures.[24] The

[18] For more information on the importance of national indicators for measuring problems, see U.S. General Accounting Office, *Forum on Key National Indicators: Assessing the Nation's Position and Progress* (GAO-03-672SP, May 2003).

[19] See Section 1005 of the Anti-Drug Abuse Act of 1988, P.L. 100-690 (Nov. 18, 1988).

[20] See P.L. 107-296, sec. 302(2).

[21] See P.L. 103-62, sec. 3.

[22] See Presidential Decision Directive 63, Critical Infrastructure Protection, May 22, 1998.

[23] Office of Management & Budget, *The President's Management Agenda, Fiscal Year 2002*, p. 29

[24] See U.S. General Accounting Office, *Combating Terrorism: Intergovernmental Partnership in a National Strategy to Enhance State and Local Preparedness*, GAO-02-547T (Washington, D.C.: March 22, 2002), p. 3, and GAO-03-519T, p. 17.

Gilmore Commission, Brookings Institution and ANSER Institute for Homeland Security also commented on the need for setting priorities (goals), measurable outcomes and assessment of activities toward these ends.

Resources, Investments, and Risk Management

This characteristic addresses what the strategy will cost, the sources and types of resources and investments needed, and where those resources and investments should be targeted. Ideally, a strategy would also identify appropriate mechanisms to allocate resources, such as grants, in-kind services, loans, and user fees, based on identified needs. Alternatively, a strategy might identify appropriate "tools of government," such as regulations, tax incentives, and standards, to mandate or stimulate nonfederal organizations to use their unique resources. Furthermore, a national strategy might elaborate on the risk assessment mentioned earlier and give guidance to implementing parties to manage their resources and investments accordingly—and begin to address the difficult but critical issues about who pays, and how such efforts will be funded and sustained in the future. Furthermore, a strategy might include a discussion of the type of resources required, such as budgetary, human capital, information, information technology (IT), research and development (R&D), procurement of equipment, or contract services. A national strategy might also discuss linkages to other resource documents, such as federal agency budgets or human capital, IT, R&D, and acquisition strategies. Finally, a national strategy might also discuss in greater detail how risk management will aid implementing parties in prioritizing and allocating resources, including how this approach will create society-wide benefits and balance these with society-wide costs. Related to this, a national strategy might discuss the economic principle of risk-adjusted return on resources.

In similar fashion, we found support for this characteristic in legislation mandating the *Money Laundering* and *National Security* strategies. Additionally, this characteristic receives related legislative and executive support, and is further supported by GAO and research policy organization publications. The *Money Laundering* strategy legislation requires a 3-year projection for program and budget priorities and a "complete assessment" of how the proposed budget is intended to satisfy strategy implementation.[25] The *National Security* strategy legislation requires an evaluation of whether the nation's "capabilities" (political, economic, and

[25] 31 U.S.C. 5341(b)(6), (7).

military) are adequate to support the implementation process.[26] Related legislative and executive branch supporting guidance for this characteristic derives from: the budget and resource balance provisions of the *National Drug Control Strategy*, HSPD-8 provisions targeting resource priorities against perceived risk of attack;[27] and the integration of performance monitoring and budgetary decision-making in the President's Management Agenda of Fiscal Year 2002.[28] GAO has also discussed the importance of this characteristic in recent testimonies, suggesting that the executive branch should link resources to threats, using a risk management approach and that carefully constructed investment strategies are needed to make appropriate use of limited fiscal and human resources.[29] The Hart-Rudman Commission and the Gilmore Commission have similarly discussed the need for a homeland security strategy to be appropriately resourced;[30] ANSER likewise has indicated the need for a strategy to be supported by a comprehensive budget plan that aligns resources with national priorities.[31]

[26] 50 U.S.C. 404a(b)(3), (4).

[27] Homeland Security Presidential Directive/HSPD-8, National Preparedness, sec. (6), Dec. 17, 2003.

[28] Office of Management & Budget, *The President's Management Agenda, Fiscal Year 2002*, p. 29.

[29] See U.S. General Accounting Office, *National Preparedness: Integration of Federal, State, Local, and Private Sector Efforts is Critical to an Effective National Strategy for Homeland Security*, GAO-02-621T (Washington, D.C.: April 11, 2002), p. 3; and GAO-03-519T, pp. 7-8.

[30] The U.S. Commission on National Security/21st Century (aka The Hart-Rudman Commission), *Seeking A National Strategy: A Concert for Preserving Security and Promoting Freedom: Phase II Report* (Ap. 15, 2000), p. 16; Second Annual Report to The President and The Congress Of the Advisory Panel to Assess Domestic Response Capabilities For Terrorism Involving Weapons Of Mass Destruction (aka Gilmore Commission), *II. Toward A National Strategy For Combating Terrorism* (Dec. 15, 2000), pp. iv, 5; Fourth Annual Report to the President and the Congress of the Advisory Panel to Assess Domestic Response Capabilities for Terrorism Involving Weapons of Mass Destruction (aka Gilmore Commission), *IV. Implementing the National Strategy* (Dec.15, 2002), p. 37.

[31] Ruth David, *Homeland Security: Building a National Strategy*, The Bridge, 32, 1 (Spring, 2002), p. 3; David McIntyre, *The National Strategy for Homeland Security: Finding the Path Among the Trees*, ANSER Institute for Homeland Security, (July 19, 2002), pp. 4-5.

Organizational Roles, Responsibilities, and Coordination

This characteristic addresses what organizations will implement the strategy, their roles and responsibilities, and mechanisms for coordinating their efforts. It helps to answer the fundamental question about who is in charge, not only during times of crisis, but also during all phases of homeland security efforts: prevention, vulnerability reduction, and response and recovery. This characteristic entails identifying the specific federal departments, agencies, or offices involved and, where appropriate, the different sectors, such as state, local, private, or international sectors. A strategy would ideally clarify implementing organizations' relationships in terms of leading, supporting, and partnering.[32] In addition, a strategy should describe the organizations that will provide the overall framework for accountability and oversight, such as the Homeland Security Council, OMB, Congress, or other organizations. Furthermore, a strategy might also identify specific processes for coordination and collaboration between sectors and organizations—and address how any conflicts would be resolved.

We found support for this characteristic in the *Money Laundering* strategy legislation, which provides that the strategy must address the coordination of regulatory and enforcement efforts; the enhancement of cooperation between federal, state, and local officials, as well as private sector entities; and the improvement of communications systems.[33] This characteristic also enjoys broad support from related legislation, executive orders, presidential directives, and recent GAO and policy research organization publications. For example, the Homeland Security Act of 2002 charges DHS with various functions, including coordination with nonfederal entities and promotion of public-private partnerships, among other things.[34] In addition, the statute mandating the *National Drug Control Strategy* calls for cooperative efforts between federal, state, and local governments and private sector initiatives.[35] Furthermore, HSPD-6, HSPD-7, PPD 63, and National Security Decision Directive (NSDD) 207 each seek to delineate the roles and responsibilities of various federal agencies and department heads; and Executive Order 13228 and HSPD-1 seek to

[32] By "partnering," we refer to shared, or joint, responsibilities among implementing parties where there is otherwise no clear or established hierarchy of lead and support functions.

[33] 31 U.S.C. 5341(b)(2), (4), (5) and (11).

[34] See P.L. 107-296, sec. 102(c), (f).

[35] Anti-Drug Abuse Act of 1988, P.L. 100-690, sec. 1005(b)(2).

coordinate implementation of the national strategy.[36] In addition, we emphasized that a national strategy should define the roles of federal, state, and local governments as well as the private sector, and that a national strategy needs to provide both direction and guidance to governments and the private sector so that missions and contributions can be more appropriately coordinated.[37] The Gilmore Commission, ANSER, and the Brookings Institution have also discussed the need for clearly assigning roles, responsibilities, accountability, liaison, and coordination among intergovernmental agencies, multilateral institutions, and international organizations.[38]

Integration and Implementation

This characteristic addresses both how a national strategy relates to other strategies' goals, objectives, and activities (horizontal integration)—and to subordinate levels of government and other organizations and their plans to implement the strategy (vertical integration). For example, a national strategy could discuss how its scope complements, expands upon, or overlaps with other national strategies. Similarly, related strategies could highlight their common or shared goals, subordinate objectives, and activities. In addition, a national strategy could address its relationship with relevant documents from implementing organizations, such as the strategic plans, annual performance plans, or annual performance reports GPRA requires of federal agencies. A strategy might also discuss, as appropriate, various strategies and plans produced by the state, local, private or international sectors. A strategy could also provide guidance such as the development of national standards to link together more

[36] See generally Homeland Security Presidential Directive/HSPD-6, Integration and Use of Screening Information, Sept. 16, 2003; Homeland Security Presidential Directive/HSPD-7, Critical Infrastructure Identification, Prioritization, and Protection, Dec. 17, 2003; Presidential Decision Directive/NSC-63, Critical Infrastructure Protection, May 22, 1998; National Security Decision Directive/NSDD-207, The National Program for Combating Terrorism, Jan. 20, 1986; Executive Order 13228, Establishing the Office of Homeland Security and the Homeland Security Council, Oct. 8, 2001; and Homeland Security Presidential Directive/HSPD-1, Organization and Operation of the Homeland Security Council, Oct. 29, 2001.

[37] See GAO-03-519T, pp. 15-16; and GAO-02-621T, p. 3.

[38] First Annual Report to The President and The Congress of the Advisory Panel To Assess Domestic Response Capabilities For Terrorism Involving Weapons of Mass Destruction (aka Gilmore Commission), *I: Assessing the Threat* (December 15, 1999), pp. x-xi; Ruth David, *Homeland Security: Building a National Strategy*, The Bridge, 32,1 (Spring, 2002), p. 5; Michael E. O'Hanlon et al., *Protecting the American Homeland: One Year On*, Brookings Institution, 2003, p. xxv.

effectively the roles, responsibilities, and capabilities of the implementing parties.

We found support for this characteristic in the *Money Laundering* strategy legislation, which requires the strategy to address how to enhance intergovernmental cooperation and the flow of information between federal, state, and local governments; the coordination of regulatory and enforcement efforts; and the role of the private sector in a more integrated approach.[39] Related legislative and executive support derives from the *National Drug Control Strategy* legislation, presidential directive and executive order. The *National Drug Control Strategy* statutory requirements call for improving the timely flow of information to federal agencies by enhancing the compatibility of automated information and communication systems.[40] In addition, HSPD-7 addresses coordination and integration,[41] and Executive Order 13228 states that executive departments and agencies shall, to the extent permitted by law, make available to the Homeland Security Council all necessary information relating to terrorist threats and activities within the United States.[42] We indicated that the national strategy would benefit from addressing how intergovernmental and private sector initiatives can be operationally coordinated and integrated and, specifically, that an "overarching, integrated framework" can help deal with issues of potential duplication, overlap and conflict.[43] Similarly, the Gilmore Commission defined a "New Normalcy" of vertical and horizontal information and intelligence sharing and ANSER has called for federal program integration where possible.[44]

[39] 31 U.S.C. 5341(b)(4), (5), and (11).

[40] Anti-Drug Abuse Act of 1988, P.L. 100-690, sec. 1005(b)(6).

[41] *See* Homeland Security Presidential Directive/HSPD-7, Critical Infrastructure Identification, Prioritization, and Protection, Dec. 17, 2003.

[42] Executive Order 13228, Establishing the Office of Homeland Security and the Homeland Security Council, sec. 3(b)(ii), Oct. 8, 2001.

[43] See GAO-02-1122T, p. 12; and GAO-03-260, p. 38.

[44] Fifth Annual Report to The President and The Congress of the Advisory Panel To Assess Domestic Response Capabilities For Terrorism Involving Weapons of Mass Destruction (aka Gilmore Commission), *V: Forging America's New Normalcy*, December, 15, 2003, pp. i, iv; David McIntyre, *the National Strategy for Homeland Security: Finding the Path Among the Trees*, The ANSER Institute for Homeland Security, 2002, p. 7.

Applying the Desirable Characteristics to the National Strategies

After developing the characteristics, we reviewed the content of each national strategy to determine the extent to which it satisfied each of the six desirable characteristics. We did this by first summarizing the structure of each strategy in terms of its overall goals, subordinate objectives, and specific initiatives. Next, we carefully read through each strategy to apply our characteristics and recorded our results on individual matrixes so we could compare characteristics across the strategies. Finally, we summarized our results on a matrix "snapshot," using our judgment to rate each national strategy on each characteristic. Strategies could obtain one of three potential scores: "addresses," "partially addresses" or "does not address." Per our methodology, a strategy "addresses," a characteristic when it explicitly cites all elements of a characteristic, even if it lacks specificity and details and thus could be improved upon. A strategy "partially addresses" a characteristic when it explicitly cites some, but not all elements of a characteristic. Within our designation of "partially addresses" there is a wide variation between a strategy that addresses most of the elements of a characteristic and a strategy that addresses few of the elements of a characteristic. A strategy "does not address" a characteristic when it does not explicitly cite or discuss any elements of a characteristic, and/or any implicit references are either too vague or general.

To verify our work, the members of the project team independently reviewed the matrix summaries at every stage and made adjustments accordingly. Specifically, the project team verified that examples of where strategies "address" or "partially address" characteristics were valid and, furthermore, that we properly characterized the strategies as not addressing the characteristics. In addition, we asked other internal teams who are familiar with the strategies from past reports and testimonies to verify our summary analysis.

GAO Related Products

Management (including Intergovernmental Coordination, Fiscal & Strategic Planning)

Terrorist Financing: U.S. Agencies Should More Systematically Assess the Use of Alternative Financing Mechanisms. GAO-04-163. Washington, D.C.: November 14, 2003.

Combating Money Laundering: Opportunities Exist to Improve the National Strategy. GAO-03-813. Washington, D.C.: September 26, 2003.

Combating Terrorism: Interagency Framework and Agency Programs to Address Overseas Threat. GAO-03-165. Washington, D.C.: May 23, 2003.

Combating Terrorism: Observations on National Strategies Related to Terrorism. GAO-03-519T. Washington, D.C.: March 3, 2003.

Major Management Challenges and Program Risks: Department of Homeland Security. GAO-03-102. Washington, D.C.: January 1, 2003.

Homeland Security: Management Challenges Facing Federal Leadership. GAO-03-260. Washington, D.C.: December 20, 2002.

Homeland Security: Information Technology Funding and Associated Management Issues. GAO-03-250. Washington, D.C.: December 13, 2002.

Combating Terrorism: Funding Data Reported to Congress Should Be Improved. GAO-03-170. Washington, D.C.: November 26, 2002.

Homeland Security: Effective Intergovernmental Coordination is Key to Success. GAO-02-1013T. Washington, D.C.: August 23, 2002.

Homeland Security: Critical Design and Implementation Issues. GAO-02-957T. Washington, D.C.: July 17, 2002.

Homeland Security: Proposal for Cabinet Agency has Merit, But Implementation Will be Pivotal to Success. GAO-02-886T. Washington, D.C.: June 25, 2002.

Homeland Security: Key Elements to Unify Efforts Are Underway but Uncertainty Remains. GAO-02-610. Washington, D.C.: June 7, 2002.

Homeland Security: Responsibility and Accountability for Achieving National Goals. GAO-02-627T. Washington, D.C.: April 11, 2002.

Homeland Security: Challenges and Strategies in Addressing Short-and Long-Term National Needs. GAO-02-160T. Washington, D.C.: November 7, 2001.

Homeland Security: A Risk Management Approach Can Guide Preparedness Efforts. GAO-02-208T. Washington, D.C.: October 31, 2001.

Homeland Security: A Framework for Addressing the Nation's Issues. GAO-01-1158T. Washington, D.C.: September 21, 2001.

Combating Terrorism: Selected Challenges and Related Recommendations. GAO-01-822. Washington, D.C.: September 20, 2001.

Combating Terrorism: Linking Threats to Strategies and Resources. GAO/T-NSIAD-00-218. Washington, D.C.: July 26, 2000.

Combating Terrorism: How Five Countries Are Organized to Combat Terrorism. GAO/NSIAD-00-85. Washington, D.C.: April 7, 2000.

Emergency Preparedness and Response

Bioterrorism: A Threat to Agriculture and the Food Supply. GAO-04-259T. Washington, D.C.: November 19, 2003.

Homeland Security: Challenges in Achieving Interoperable Communications for First Responders. GAO-04-231T. Washington, D.C.: November 6, 2003.

September 11: Overview of Federal Disaster to the New York City Area. GAO-04-72. Washington, D.C.: October 31, 2003.

Homeland Security: Reforming Federal Grants to Better Meet Outstanding Needs. GAO-03-1146T. Washington, D.C.: September 3, 2003.

Hospital Preparedness: Most Urban Hospitals Have Emergency Plans but Lack Certain Capacities for Bioterrorism Response. GAO-03-924. Washington, D.C.: August 6, 2003.

Bioterrorism: Information Technology Strategy Could Strengthen Federal Agencies' Abilities to Respond to Public Health Emergencies. GAO-03-139. Washington, D.C.: May 30, 2003.

Bioterrorism: Adequacy of Preparedness Varies Across State and local Jurisdictions. GAO-03-373. Washington, D.C.: April 7, 2003.

Homeland Security: Intergovernmental Coordination and Partnerships Will Be Critical to Success. GAO-02-899T. Washington, D.C.: July 1, 2002.

National Preparedness: Integration of Federal, State, Local, and Private Sector Efforts is Critical to an Effective National Strategy for Homeland Security. GAO-02-621T. Washington, D.C.: April 11, 2002.

Combating Terrorism: Enhancing Partnerships Through a National Preparedness Strategy. GAO-02-549T. Washington, D.C.: March 28, 2002.

Combating Terrorism: Critical Components of a National Strategy to Enhance State and Local Preparedness. GAO-02-548T. Washington, D.C.: March 25, 2002.

Combating Terrorism: Intergovernmental Partnership in a National Strategy to Enhance State and Local Preparedness. GAO-02-547T. Washington, D.C.: March 22, 2002.

Homeland Security: Progress Made; More Direction and Partnership Sought. GAO-02490T. Washington, D.C.: March 12, 2002.

Combating Terrorism: Key Aspects of a National Strategy to Enhance State and Local Preparedness. GAO-02-473T. Washington, D.C.: March 1, 2002.

Combating Terrorism: Considerations for Investing Resources in Chemical and Biological Preparedness. GAO-02-162T. Washington, D.C.: October 17, 2001.

Bioterrorism: Review of Public Health and Medical Preparedness Programs. GAO-02-149T. Washington, D.C.: October 10, 2001.

Combating Terrorism: Observations on Options to Improve the Federal Response. GAO-01-660T. Washington, D.C.: April 24, 2001.

Combating Terrorism: Issues in Managing Counterterrorist Programs. GAO/T-NSIAD-00-145. Washington, D.C.: April 6, 2000.

Border and Transportation Security

Homeland Security: Preliminary Observations on Efforts to Target Security Inspections of Cargo Containers. GAO-04-325T. Washington, D.C.: December 16, 2003.

Aviation Security: Efforts to Measure Effectiveness and Strengthen Security Programs. GAO-04-285T. Washington, D.C.: November 20, 2003.

Aviation Security: Efforts to Measure Effectiveness and Address Challenges. GAO-04-232T. Washington, D.C.: November 5, 2003.

Homeland Security: Overstay Tracking is a Key Component of a Layered Defense. GAO-04-170T. Washington, D.C.: October 16, 2003.

Coast Guard: New Communication System to Search and Rescue Faces Challenges. GAO-03-1111. Washington, D.C.: September 30, 2003.

Airport Passenger Screening: Preliminary Observations on Progress Made and Challenges Remaining. GAO-03-1173. Washington, D.C.: September 24, 2003.

Homeland Security: Risks Facing Key Border and Transportation Security Program Need to be Addressed. GAO-03-1083. Washington, D.C.: September 19, 2003.

Maritime Security: Progress Made in Implementing Maritime Transportation Security Act, but Concerns Remain. GAO-03-1155T. Washington, D.C.: September 9, 2003.

Transportation Security: Federal Action Needed to Enhance Security Efforts. GAO-03-1154T. Washington, D.C.: September 9, 2003.

Aviation Security: Progress Since September 11th and the Challenges Ahead. GAO-03-1150T. Washington, D.C.: September 9, 2003.

Land Border Ports of Entry: Vulnerabilities and Inefficiencies in the Inspections Process. GAO-03-1084R. Washington, D.C.: August 18, 2003.

Container Security: Expansion of Key Customs Programs Will Require Greater Attention to Critical Success Factors. GAO-03-770. Washington, D.C.: July 25, 2003.

Border Security: New Policies and Increased Interagency Coordination Needed to Improve Visa Process. GAO-03-1013T. Washington, D.C.: July 15, 2003.

Transportation Security: More Federal Coordination Needed to Help Address Security Challenges. GAO-03-843. Washington, D.C.: June 30, 2003.

Homeland Security: Challenges Facing the Department of Homeland Security in Balancing Its Trade Facilitation and Border Protection Missions. GAO-03-902T. Washington, D.C.: June 16, 2003.

Transportation Security: Post 9/11 Initiatives and Long-Term Challenges. GAO-03-616T. Washington, D.C.: April 1, 2003.

Border Security: Challenges in Implementing Border Technology. GAO-03-546T. Washington, D.C.: March 12, 2003.

Aviation Security: Vulnerabilities and Potential Improvements for the Air Cargo Security System. GAO-03-344. Washington, D.C.: December 20, 2002.

Port Security: Nation Faces Formidable Challenges in Making New Initiatives Successful. GAO-02-993T. Washington, D.C.: August 5, 2002.

Information Analysis and Infrastructure Protection

Critical Infrastructure Protection: Challenges in Securing Control Systems. GAO-04-140T. Washington, D.C.: October 1, 2003.

Homeland Security: Information Sharing Responsibilities, Challenges, and Key Management Issues. GAO-03-1165T. Washington, D.C.: September 17, 2003.

Homeland Security: Counterfeit Identification and Identification Fraud Raise Security Concerns. GAO-03-1147T. Washington, D.C.: September 9, 2003.

Homeland Security: Efforts to Improve Information Sharing Need to Be Strengthened. GAO-03-760. Washington, D.C.: August 27, 2003.

Homeland Security: Information Sharing Responsibilities, Challenges and Key Management Issues. GAO-03-715T. Washington, D.C.: May 8, 2003.

Information Technology: Terrorist Watch Lists Should Be Consolidated to Promote Better Integration and Sharing. GAO-03-322. Washington, D.C.: April 15, 2003.

Critical Infrastructure Protection: Challenges for Selected Agencies and Industry Sectors. GAO-03-233. Washington, D.C.: February 28, 2003.

Protecting Information Systems Supporting the Federal Government and the Nation's Critical Infrastructure. GAO-03-121. Washington, D.C.: January 30, 2003.

Homeland Security: Information Sharing Activities Face Continued Management Challenges. GAO-02-1122T. Washington, D.C.: October 1, 2002.

National Preparedness: Technology and Information Sharing Challenges. GAO-02-1048R. Washington, D.C.: August 30, 2002.

Critical Infrastructure Protection: Federal Efforts Require a More Coordinated and Comprehensive Approach to Protecting Information Systems. GAO-02-474. Washington, D.C.: July 15, 2002.

Homeland Security: Key Elements of a Risk Management Approach. GAO-02-150T. Washington, D.C.: October 12, 2001.

Science and Technology; Chemical, Biological, Radiological, and Nuclear Countermeasures

Nuclear Security: Federal and State Action Needed to Improve Security of Sealed Radioactive Sources. GAO-03-804. Washington, D.C.: August 6, 2003.

Nuclear Regulatory Commission: Oversight of Security at Commercial Nuclear Power Plants Needs to be Strengthened. GAO-03-752. Washington, D.C.: September 4, 2003.

Nuclear Nonproliferation: U.S. and International Assistance Efforts to Control Sealed Radioactive Sources Need Strengthening. GAO-03-638. Washington, D.C.: May 16, 2003.

Homeland Security: Title III of the Homeland Security Act of 2002. GAO-02-927T. Washington, D.C.: July 9, 2002.

GAO's Mission

The General Accounting Office, the audit, evaluation and investigative arm of Congress, exists to support Congress in meeting its constitutional responsibilities and to help improve the performance and accountability of the federal government for the American people. GAO examines the use of public funds; evaluates federal programs and policies; and provides analyses, recommendations, and other assistance to help Congress make informed oversight, policy, and funding decisions. GAO's commitment to good government is reflected in its core values of accountability, integrity, and reliability.

Obtaining Copies of GAO Reports and Testimony

The fastest and easiest way to obtain copies of GAO documents at no cost is through the Internet. GAO's Web site (www.gao.gov) contains abstracts and full-text files of current reports and testimony and an expanding archive of older products. The Web site features a search engine to help you locate documents using key words and phrases. You can print these documents in their entirety, including charts and other graphics.

Each day, GAO issues a list of newly released reports, testimony, and correspondence. GAO posts this list, known as "Today's Reports," on its Web site daily. The list contains links to the full-text document files. To have GAO e-mail this list to you every afternoon, go to www.gao.gov and select "Subscribe to e-mail alerts" under the "Order GAO Products" heading.

Order by Mail or Phone

The first copy of each printed report is free. Additional copies are $2 each. A check or money order should be made out to the Superintendent of Documents. GAO also accepts VISA and Mastercard. Orders for 100 or more copies mailed to a single address are discounted 25 percent. Orders should be sent to:

U.S. General Accounting Office
441 G Street NW, Room LM
Washington, D.C. 20548

To order by Phone:	Voice:	(202) 512-6000
	TDD:	(202) 512-2537
	Fax:	(202) 512-6061

To Report Fraud, Waste, and Abuse in Federal Programs

Contact:

Web site: www.gao.gov/fraudnet/fraudnet.htm
E-mail: fraudnet@gao.gov
Automated answering system: (800) 424-5454 or (202) 512-7470

Public Affairs

Jeff Nelligan, Managing Director, NelliganJ@gao.gov (202) 512-4800
U.S. General Accounting Office, 441 G Street NW, Room 7149
Washington, D.C. 20548

PRINTED ON RECYCLED PAPER

Mr. SHAYS. We'll start with Congressman Schrock first; and then we'll go to you, Mr. Ruppersberger, and then to you, Mr. Vice Chairman, and then to you, John.

Mr. SCHROCK. Thank you, Mr. Chairman.

Thank you, Mr. Yim. Fascinating remarks.

I, too, worry about complacency. Every day we get further away from September 11, I worry more and more.

This is your matrix?

Mr. YIM. Yes, sir.

Mr. SCHROCK. I was fascinated by those that were mixed, mediocre, or weak; and that's not good. This certainly needs to be improved.

I don't know how quick it's going to happen, and the desired objective is not linked to money. That seems to be the key to everything up here. It seems we have to put our money where our objectives are or we're going to pay for it.

I'm going to make a couple comments, and I'm going to let the second panel know we are going to ask the same questions.

I believe the National Security Strategy is a forward-looking vision that goes a long way toward reorienting our Nation toward the post-September 11 world.

I do note as a document focused primarily on international relations, reorienting military and intelligence capabilities is only mentioned in a cursory fashion. While a companion national military strategy has been written, I'm not aware of a similar national intelligence strategy.

Though there is no doubt in my mind that we possess the finest military and intelligence capabilities in the world, I remain uneasy about our ability to evaluate non-traditional and asymmetric threats and to integrate the many different strains of intelligence that we gather.

That being said, in your opinion, should we develop a national intelligence strategy that addresses these perceived weaknesses; and, if developed, what would you recommend such strategy address?

You touched on some of that, but I wonder if you could go into more detail.

Mr. YIM. Yes, that is certainly a key issue. Threat and risk assessment based on good intelligence is a critical precursor to setting our priorities and allocating the resources effectively and cost effectively. While most of our criteria that we discussed today talked about transparency and accountability, there will be a need for secrecy in a national intelligence strategy. On the other hand, more and more people need to be connected to the intelligence communities that have not been in the past; and those people are unfamiliar as to who to call, what expectations on the type of information that they will receive, the detailed nature of that information. So I think that makes it all imperative that we have some sort of national strategy.

Generally, some of the national strategies do discuss intelligence issues. For example, the National Homeland Security Strategy has a primary section on intelligence and warning, talking about building new capabilities through the information assurance and information infrastructure profession directorate.

The Combating Terrorism Strategy talks about locating terrorists in their organization and assessing intelligence capabilities to gather human and technical intelligence.

The Combating Terrorism Strategy also references this TTIC, the Terrorist Threat Integration Center, and talks about the need for intelligence fusion, taking all of the data that's being gathered by our intelligence community and fusing it into adequate material. This is in various locations.

Does it need to be brought together? I think that's one of the various purposes of the Terrorist Threat Integration Center; and I think our discussion from the State, local and private sector, they would like a more coordinated way to receive threat information so they can plan accordingly.

Mr. SCHROCK. I agree.

You said who to call? I think somebody told me there were 47 Federal agencies that did intelligence after September 11. Nobody would share with anybody, and I think that's a big problem. God forbid we suggest merge the CIA and FBI together. There would be a revolt like you've never seen, but it's coming.

No. 2, the strategies that GAO submitted reporting to this committee state that—in an unequivocal fashion our national policies toward a variety of threats from both traditional and non-traditional actors. Our goals are clearly stated.

As leaders, we have become comfortable with the idea that the war on terror must be a sustained and lasting effort. We believe we must not use that fact as an excuse to prolong our evaluation of our short-term progress. These strategies for the most part do not include metrics or milestones to be used to measure our progress.

Question: Should we develop a timeline along which to assess our progress in implementing these strategies, and what would you propose as metric suitable for measuring our progress toward achieving the stated goals and objectives of these strategies?

Mr. YIM. Yes, I think, sir, that the timelines are imperative. People do react to timelines.

I think initially when the strategies were developed, because so much needed to be done and it wasn't clear how we were going to approach some of these issues, that, in fairness, some of them did not have timelines.

However, we have seen iterations now. Further documents come out from our national statutes. We've had firm timelines imposed by the Congress on baggage screening, for example. We've had firm deadlines imposed upon the Coast Guard for port vulnerability assessments. We recently had the administration issue two Presidential directives, Homeland Security Directives No. 7 and 8, in December 2003, that assigned firm responsibilities and tasked the secretaries of the responsible Cabinet agencies within fixed periods of time to develop certain strategies, to develop performance metrics. I think that's clearly what we need.

What Congress has done in certain areas is legislate or mandate particular timeframes, and I think that may be an option that could be considered.

One of the dangers is that sometimes that may tie or limit some of the flexibility, but certainly I think for the Congress to exert that type of oversight is certainly something that should be considered, sir.

Mr. SHAYS. Thank you.

Let me just say that we usually go 10 minutes. I think with so many Members we probably should do a first round of 5 minutes.

I just want to say, for all the Members, the first hearing we had was a hearing that said we had no real strategies. Now we have this proliferation of eight strategies; and this hearing is to kind of evaluate how we're doing on these strategies and what is, in essence, a good strategy, how do we determine that.

And at the third hearing we're going to have—I just want to put it on the record, Mr. Ruppersberger, because you mentioned it—the third hearing we will have government witnesses. The administration needs to come and say, OK, we know we went from none to many and now we're trying to evaluate them and this is what we're finding. What's your response and where are we.

Mr Ruppersberger, I recognize you.

Mr. RUPPERSBERGER. Mr. Yim, you stated there was considerable variation to the extent of the strategies and how it related to homeland security and terrorism and that all the strategies identified goals, supported objectives, and other characteristics. But the strategies generally, from what I'm hearing from your testimony and correct me if I'm wrong, do not address resources, investments, and risk management, or integration, implementation. And even where the characteristics are addressed, improvements could be made.

For example, while the strategies identify goals, support objectives and specific activities, they generally do not address or discuss priorities, milestones, or performance measures, which is where we want to get, where our goal is; and the elements are desirable for evaluating progress and achieving oversight.

Now you stated the strategies range from strong to weak in defining problems. For example, Homeland Security, Cyberspace and Critical Infrastructure Strategies were judged to be the most developed, while National Security Strategy and WMD were considered to be the most vague and weakest; is that correct?

Mr. YIM. That's correct, sir.

Mr. RUPPERSBERGER. Now do different levels of maturation and subject expertise really account for all the differences?

Mr. YIM. I think that accounts for some of the differences but not all of the differences.

The value, as I said, of our analysis is it is comparative analysis. You could expect the National Security Strategy, the most top-level strategy, would probably be the most general one in nature. You would expect the Money Laundering Strategy, which is targeted for specific agencies—FBI, law enforcement—that has a long history of criminal activities would be more definite in defining roles and responsibilities.

Mr. RUPPERSBERGER. WMDs have been around longer than cyberspace.

Mr. YIM. Yes, and I think that really talks about counterproliferation, nonproliferation and just management in very general terms; and it doesn't—is it useful in such general terms for people that are going to be charged with implementation of the strategy? I think that's the question we're raising.

Other strategies like the National Security Strategy, the Homeland Security Strategy had ways to be specific. They said who was in charge of specific activities. Perhaps that could be added to the Weapons of Mass Destruction Strategy. Perhaps there could be some timelines added to the WMD strategies.

Performance measures? That's perhaps hard to judge.

So we're not saying that each strategy has to be at the same level, but I think there's significant lessons from each strategy. And each could be improved, all could be improved, of course.

Mr. RUPPERSBERGER. Why do you think the administration really outlined the strategies the way they did?

Mr. YIM. It's hard for me to speculate.

I think one of the reasons some of the strategies are less specific is that, in certain areas, so much needed to be done right after September 11 that even general strategies were useful to mobilize the resources.

We were so lacking in preparedness, despite the Bremer Commission, Gilmore Commission, Hart-Rudman Commission recommendations, that immediately after the September 11 attacks just focusing attention on certain key areas was a useful exercise for the Nation. I think the need to get a strategy out quickly to mobilize the support was a good goal of the administration, but we're beyond that now.

We're, as Mr. Schrock indicates, at a danger of complacency. We need to move toward the implementation stage. And that means the strategies have to firm up, they have to get sharper; and until we do that and provide some performance measurements we're not going to know whether the commitment of resources is really making it safer.

Mr. RUPPERSBERGER. It could have been because of September 11 that the strategies were hastily written by the administration, in all fairness to them, because there was none before that, correct?

Mr. YIM. I really am not——

Mr. RUPPERSBERGER. We're only trying to get to the end game, and that's the purpose.

Do you feel, though, when we're dealing with strategies in these issues and especially such national strategies that before we come out with the strategies that we deal with the facts and data and get more data to come with a more concrete strategy than the way it is now?

Mr. YIM. I think that's exactly right, sir.

We do need now to move. When we move with implementation, it has to be supported with good data. That's not only data on risks and threats and intelligence data but on our infrastructure.

Do we really know what our hospital infrastructure is capable of doing for a SARS attack or an avian bird flu virus?

Do we really know what our power grids can do under certain situations, not only an attack but a human error that led to the cascading Northeast blackouts, and how quickly can they recover if there was an exerted—worm or virus being exerted into the system?

We need better data. When I was in the Department of Defense, one of the things that really hindered us in doing our infrastructure recapitalization was a fundamental lack of data available. We didn't really know what we owned and what we controlled, and if you don't know that information—and in many senses we don't know exactly what the capabilities of the State and local and private sector are to respond or to be prepared in certain areas—it's very difficult to develop a strategy and to implement.

Mr. RUPPERSBERGER. And the local and State issue is a major issue, also, in bringing them all together?

Mr. YIM. Yes.

Mr. RUPPERSBERGER. Thank you, Mr. Chairman.

Mr. SHAYS. Thank you.

At this time I recognize Mr. Turner.

Mr. TURNER. Thank you, Mr. Chairman.

I would echo Mr. Schrock's statement with respect to a national strategy on intelligence. Because certainly in reading the description of the various strategies, intelligence comes out in each of them; and as we talk about first responders and to agencies and, of course, agencies that are responsible for intelligence gathering, the coordinated effort both in gathering and dissemination of intelligence is really probably the most important aspect of our preparedness with respect to combating terrorism.

You spoke about the issue of the strategies themselves and the lack of definitive information on the implementation for agencies and that—really looking at various strategies and the lanes they're in and how really, going forward, each agency might implement aspects of them. I'm interested in the coordination between strategies and agencies, to what extent the strategies provide guidance or to what extent the agencies are looking at the various strategies before them, coordinating their implementation of the strategies or even the agency's efforts with other agencies.

Mr. YIM. That issue of horizontal integration among the Federal agencies is critical.

When many Federal agencies look at strategies, they talk about their obligations under the GPRA-type of requirements. That's very narrowly agency focused. When we're talking about the Homeland Security Strategies, we're talking about issues that cross-cut over and above a particular agency's jurisdiction. When we are talking about preparedness for a bioterrorism event, it's not only HHS, it's DHS, it's going to be Justice, it's going to be DOD, it's going to be a variety of other agencies. So the key is the strategy would have to cross-cut the agency jurisdictions.

Do they do that enough? We found mixed results when we talk about who's in charge. Sometimes they talk about lead agencies, but sometimes they do not. Sometimes they don't add the time component. There may be a lead agency for prevention but a different agency for response or recoverability assessment.

I think it's illustrative to look at the Homeland Security Presidential Directives that came out in December. I think they responded to some of the criticisms about the national strategy, and they were very specific. They said, you, Secretary of HHS, you, Secretary of DHS, you, Secretary of Energy, are to do these specific things, and you are to coordinate your activities in this specific manner, but the overall lead is "X."

I think that is a good example of where we would like some of these strategies to head, because I think we have to recognize they cross-cut well beyond the ability of any single Federal agency and even the Federal Government.

We need to talk about vertical integration. The Feds can't do it all. State and locals are going to have to do stuff.

The private sector owns 80 percent of the critical infrastructure. They are going to have to do that, too.

Mr. TURNER. You talked about the issue of feedback as relates to implementation. Is there any presence of a mechanism for inter-agency feedback, where one agency who has needs from another that's not being met has an ability to accept within their own agency—cause it to be known of the need or the lack of response or the lack of implementation?

Mr. YIM. We had raised some of our concerns about that.

The Department of Homeland Security has an Interagency Coordinating Council, and they have that function. They also have a Homeland Security Advisory Council that includes State and local and private sector input to the development of their strategy, but sometimes they have to come up, butt heads, against other Cabinet agencies.

How do you prioritize homeland security against education security, energy security, hospital, health care security, and where are those balancing decisions being made, the coordination? Of course, in the executive branch, in the White House, in the Homeland Security Council, perhaps? Is that in the National Economic Council?

I think that still needs to be better clarified, and the Congress could provide I think great assistance in the balancing that needs to be occurring between very many—there are so many important priorities that this Nation has to address.

Mr. SHAYS. I thank the gentleman.

Mr. Tierney.

Mr. TIERNEY. Thank you, Mr. Yim, for your testimony and your report.

I'm concerned about what I think is an apparent failure to integrate the strategic decisionmaking between international and national criteria objectives on that. Would you speak a little bit to that?

It seems to me we have $10 million going to national missile defense, we have billions of other dollars going to weapons platforms that I think will look a little bit back toward the cold war as opposed to what we are going to do and only $1 billion in moneys allocated in port security against the possible introduction of nuclear materials in that manner. What should we do and how does this stack up in terms of international and national planning and what can we do to improve that aspect?

Mr. YIM. I think many of our strategies understandably are inwardly focused right now because of the immediate response to September 11, but clearly what we need to talk about is borderless security.

When we talk about border security, it really is a bit of an illusion. Our borders are—because of our society are designed to be free and open, to be easily passable through.

We talk about cybersecurity. There is really no sense of a border. So if we're talking about borderless security, then we clearly would need international cooperation; and strategies need to address that.

Obviously, we need to interdict a dirty bomb or a nuclear device in a cart or container before it arrives in the Port of Philadelphia or the Port of Los Angeles, and the only way we are going to get that is through the international cooperation.

Now some of the strategies address that. The Combating Terrorism Strategy talks about involving the international community. The High-Level National Security Strategy talks about, well, if we're fighting terrorism, we not only need to defeat the existing terrorism, we have to prevent the growth of new terrorists by winning the, "war of ideas."

Are we doing enough in that arena?

Well, I think some of the international community may be dismayed that we are taking unilateral actions.

Are our own protocols consistent with their business models, for example?

I think that is a fundamental purpose of going toward some type of national standards and using an international systems type standard organization that specifically factors in the considerations of the international community and the U.S. community so that they are compatible.

We depend upon foreign trade and export and import, so we must need the international cooperation for cargo security. We depend upon security, so we need the international cooperation for visas verification and terrorist watch.

So I agree with you, sir, that definitely needs to be a component of each strategy. I think in this day and age we really do not have just a homeland security strategy. It really is a global strategy.

Mr. TIERNEY. Did you see any evidence in your review of what's going on of any budgetary planning that cuts across the international and the national aspects of strategy; in other words, allocating our resources as between one and the other, going back to the example that I gave, where it seems entirely skewed?

Mr. YIM. It's difficult to see that with enough granularity, because many of the activities deal with international topics, are dual purpose or multi-purpose activities. So it's hard to split out this particular funding for increase in Department of State staff or particular programs only designed to counterterrorism. They could only be part of the overseas economic development and economic assistance programs.

So the answer is yes. If we need to have greater granularity, it's difficult in the way that the budgets are submitted to see that direct link between foreign support budgets and the counterterrorism activities.

Mr. TIERNEY. Well, would you agree with me that the prospect of having somebody bring over a dirty bomb in a container of a ship is probably far more likely than somebody getting an intercontinental ballistic missile with it targeted and directed to the United States at this point in time?

Mr. YIM. I'm sure that's correct. I'm sure the experts behind me would agree with me, also, sir.

Mr. TIERNEY. So it would seem that we concentrate more on the former than the latter in terms of how we allocate our resources. Does that make sense to you.

Mr. YIM. Absolutely. Some risk threat assessment is required.

Mr. TIERNEY. And do you see that between international and other types of threats.

Mr. YIM. Some of the strategies really only peripherally touch on that; and I think that is an area where we talked about integration, implementation.

When we talked about integration, it wasn't just in the United States. It was definitely with the international community. That was one of the major issues that we flagged during our review. We definitely need improvement in that area.

Mr. TIERNEY. Thank you.

Mr. SHAYS. I thank the gentleman.

Mr. Murphy.

Mr. MURPHY. Thank you, Mr. Chairman; and thank you, Mr. Yim.

Mr. YIM. Thank you.

Mr. MURPHY. On the issues of intelligence and coordination intelligence, certainly within intelligence agencies one of the things they also must protect is horizontal and vertical distribution of information in order to keep information secret; and yet you have to know when to distribute it horizontally and vertically in order to allow other persons to act on that.

Part of Homeland Security is to try to coordinate the efforts of FDICA, NCICA, NSA, etc. Of course, what is becoming clear in the news, too, is that many times we have—or in the last decade or so there's been depletion of perhaps agents or other folks who were able to gather active information, completely wiping out our ability from Asia, the continent of Africa and many areas in the Middle East; and we will suffer the consequences of that depletion for a while because we have not had eyes and ears on the ground. We have been relying on troop movements when we should have been looking at individuals.

Given that integration of information, one of the things I look at on a local level is the question of where do we stand now in terms of getting accurate information to all the folks who are really seen as the first and last responders on the ground—the police, the fire, the hospitals—in being able to deal with these and to have accurate information. Because I think, as we see flights canceled from Europe, as we see alerts go up and down, we certainly don't want to have the public become compliant and unresponsive, which would only increase our risk, but, nonetheless, we want to make sure that they have trust and faith in information coming through.

Where do we stand, in your assessment, on accurate information being gathered and accurate information being disseminated such as not to lead to complacency?

Mr. YIM. I think that's a very common concern that we hear voiced to us from the State and local sector, the lack of detailed information that would allow them to stay specific actions. I mean, they have been critical of the color code, the terrorist threat advisory system, in being too non-specific, that they've asked for more region specific or sector specific information. They've pointed out that they don't need to compromise sources and methods, that the cop on the street doesn't need to know how the information was gathered but only whether you want me to look up or down under the bridge, etc., on the roadways, to take effective action.

I think one of the additional concerns would be that people are unfamiliar with the intelligence community and the nature of the information that's being generated. They may lack the capabilities to analyze, certainly analyze, the raw data. So the information I think not only has to be a mechanism to provide it. They have to do some analysis to the type of information that will be provided, information that isn't going to require training to be able to analyze but information that could be actionable by a fire department chief, by a mayor, by a sheriff.

I think we are going to overcome it; and, in fact, people are going to get flooded with the data they will receive. The key will be an analysis of the data, synthesis of it, to the extent it is useful.

Mr. MURPHY. Where do we stand in the timeline of reaching that goal?

Mr. YIM. I think that has been one of the most common concerns of any—not any but one of the primary areas for additional attention and in setting some timelines for putting a plan, putting some metrics, as to what is being pumped out, getting feedback loops from State and local as to what information is most useful to them, what information they don't need. I think if we got that feedback we can overcome some of the sources and methods.

Mr. MURPHY. Are we weeks or months away from reaching that goal?

Mr. YIM. I'm not sure I'm in a position to say, sir. I wish I could. On many of these areas, I think we're more—certainly, it's more a long term than it is a short term.

Mr. MURPHY. Thank you, Mr. Chairman.

Mr. SHAYS. I was telling Mr. Ruppersberger that I'm happy I'm not in school, being tested on this; and yet I have a bit of guilt because this is so important. As one of our witnesses is going to say later, ready, fire, aim; and that's kind of what we did when we had the three commissions before us.

They said there is a threat, you need to know the threat, you need to develop the strategy, you need to organize your government. It would clearly—and this is what Mr. Tierney, frankly, on this committee has argued more than anyone else, what's their strategy?

What I am having a hard time wrestling with, and I'll kind of share some of my ignorance, which I do more often than I'd like, but when I look at the matrix, I look at seven strategies, national strategies, and I look at the national security.

Do you have that in front of you?

Mr. YIM. Yes, sir.

Mr. SHAYS. I'm going to quickly run them down.

National Security is NSC; Homeland Security, DHS, Combating Terrorism, NSC; Weapons of Mass Destruction, NSC; Physical Infrastructure, DHS; Secure Cyberspace, DHS; and Money Laundering, Treasury?

Mr. YIM. Yes, sir.

Mr. SHAYS. Now what I'm also learning from this is only National Security and Money Laundering were strategies we had developed before September 11.

Mr. YIM. Yes, sir, that's correct.

Mr. SHAYS. And what I have a sense is—yes?

Mr. YIM. I think there had been iterations of strategies. They were only published—five of the strategies were published post-September 11. Only National Security and Money Laundering were actually published in this format prior to September 11.

Mr. SHAYS. Well, I have a sense that we've gotten pretty lazy. In other words, the cold war threat we are pretty clear that it was containment, reactive nuclear destruction. What's unsettling for my constituents is that it's probably detection, prevention, maybe sometimes preemption, obviously based on better information than we had, and sometimes maybe the lateral, and I'm just talking in a general sense.

When I look at National Security, the matrix that you have, purpose, scope, and methodology it does not address problem definition and risk assessment; does not address resources investment and risk management; does not address organizational roles and responsibilities and coordination; does not address integration and implementation; does not address—and only one is partially addressed, and that's goals, objectives, activities performance measures.

Tell me why I shouldn't be hugely concerned about that.

Mr. YIM. Again, what we were trying to avoid is to give a score card, an absolute measure.

Mr. SHAYS. You don't have to give a score card on this.

Mr. YIM. But the point would be that perhaps such a high-level strategy—in fairness, in something at the top-level strategy, the National Security Strategy, it's not surprising that it would be vaguer or use more general language.

However, we still have to ask that things have changed since 1947 when we first were required to develop a National Security Strategy. Things have changed since 1988, when the statutory requirements for the National Strategy were promulgated by this Congress.

Have the world changes now, with the changing terrorist threat, the pace of technological development, required the strategies to become more specific?

I think that is our general conclusion; and I think, yes, Mr. Chairman, you should be concerned that the National Security Strategy isn't as specific as some of the other ones are.

Mr. SHAYS. Tell me this: Of the so-called desirable characteristics, which is the most important?

Mr. YIM. Well, of course, we will say that all are important or we wouldn't—but to answer your question seriously, we would say that the resource investments and the performance metrics ones are the keys.

You have to be able to sustain the effort. It's not enough to have high-level goals if the money isn't going to follow along. If people aren't going to invest the money, the people, the prioritization to achieving those objectives and if that is the most important objectives you want people to implement, then you have to have some way of telling whether or not they are spending their money correctly, and that's why the performance measurements—

Mr. SHAYS. Give me another one that's most important, second one that's way up there.

Mr. YIM. I think integration is the key. We talked about if you consider homeland security as a stand-alone item, it's going to be enormously expensive.

Mr. SHAYS. So I would have picked out goals, objectives, activities and performance measurements. That's what I would have picked as No. 1. Tell me why that doesn't top the two that you mentioned.

Mr. YIM. Again, it's difficult for me, but because I believe that where we're focused on implementation—it's a question for us—is can we afford to do everything people are identifying that needs to be done?

Mr. SHAYS. Right.

Mr. YIM. It's fine to set goals and objectives, but the reality is we are not going to be able to achieve all of those goals and objectives immediately, so how we resource and set priorities I think is going to be the key, Mr. Chairman.

Mr. SHAYS. Tell me which is the least important of that group. I mean, they're all important, but which is the least important?

Mr. YIM. I think every strategy has a general purpose statement, so in terms of utility to the user, something that says promote the common defense, ensure domestic tranquility, that's a burden of proof statement that would have been addressed in our criteria.

Mr. SHAYS. Purpose and scope?

Mr. YIM. Purpose and scope—

Mr. SHAYS. I'm sorry, is that the one you said?

Mr. YIM. That's the one I said—

Mr. SHAYS. I understand they're all important.

Then tell me—oh, jeez.

Let me just ask you: I just was verifying that I was looking at the matrix; and my staff said, "Yes." I was looking at the matrix on page 2, and then I said on figure 1 on page 8 what am I looking at? And the comment was a mess, a chart that is hard to understand.

Do you want to break this down in a Top Secret briefing or do you want to just quickly tell us what that means—a strategy of hierarchy?

Mr. YIM. When we put that graphic out, people said, you put a dunce cap in your testimony. We would prefer to consider it a wizard's cap, Mr. Chairman, but basically what we're trying to point out is that there is a hierarchy. We expect that the strategies are not the "be all and end all," that we expect the strategies to have some general statements, to be at the top of that cone or pyramid and that we expect that the responsible parties charged with implementation are going to develop further documentation, and it's going to get more granularity as you move down the cone.

So as you move from the top of the cone, the National Security Strategy, to implementing documents such as the Homeland Security Presidential Directives to specific agency strategies, that's as you move down the cone, you're getting more specificity and you would demand more performance measures.

Mr. SHAYS. Are there Members that have another question of this panel?

Mr. SCHROCK. Yes.

Mr. SHAYS. Before you go, I want you to tell me, are there any national strategies that have been left out, plus the eight which we are going to have under closed door; but if you could just start to think about what strategies should be there that are not. Ed, why don't you go. Is that all right?

Mr. SCHROCK. Just a couple quick comments, Mr. Yim. We were talking about the hierarchy could be the problem. Is the strategy the point or the mentality of the bureaucracy to put this together? You mentioned a couple of times the coordination of the agencies, the butting of the heads of Cabinet members, everybody has their own turf and nobody wants to give it up. Is that the problem?

Mr. YIM. I think that is one problem that the strategies have to address.

Mr. SCHROCK. How do we solve it?

Mr. YIM. I think that there needs to be a clear directive to our Federal agencies that they must admit that certain things are beyond their jurisdiction and scope, they must rely upon others to assist in these areas; that it is not exclusively the province—homeland security is not exclusively the province of a particular agency, a Cabinet Secretary or Department and that a topic like bioterrorism is going to overwhelm the resources of a single agency. That is the value of a strategy. How we make that realization come to pass has been a classic question.

Mr. SCHROCK. Clear directive from whom?

Mr. YIM. I think the administration clearly has that responsibility.

Mr. SCHROCK. The President of the United States?

Mr. YIM. Yes.

Mr. SCHROCK. You talked about a borderless society. We all love that. The fact is that is probably never going to happen again. How much interaction or coordination in this effort do you think needs to be made with some of our allies, some of our partners in this, a little, a whole bunch, none?

Mr. YIM. I think that is going to be a crucial aspect of it. The burden of defeating terrorism on a global scale is not going to be able to be met solely by the United States.

Mr. SHAYS. Could you say that again?

Mr. YIM. The burden of fighting terrorism on a global scale cannot be met solely by the United States. And certainly the impact of terrorists' attacks is not only felt by the United States even if the attack was only on U.S. soil. We know that 47, 50 or so countries had citizens in the World Trade Center attacks. The financial market ramifications were extended well beyond the U.S.' financial systems. So the international community being aligned in the fight against terrorism is going to be crucial.

Mr. SCHROCK. Thank you, Mr. Chairman.

Mr. SHAYS. Could I ask you on that point, I want you to say it again, and I want you to tell me under what basis you can say it. I happen to believe it, but I believe it intuitively. Is it so obvious that it stares us in the face, or is there work to be done that says categorically you can make that statement?

Mr. YIM. I think we can use examples. The cargo container security work that we have been doing, Mr. Chairman, could not be done—we can't interdict all of the cargoes without cooperation from the superports in the foreign areas telling us what is going in or having some protocols to secure who's loading what onto those containers. And containers aren't the only problem. There are great bulk carriers that are coming in, too. You could put a bomb in a grain ship as well as a container ship. When we talk about cybersecurity, we certainly know that it's not just that.

Mr. SHAYS. Thank you for your patience.

Mr. RUPPERSBERGER. I do agree with the issues.

Bottom line questions: First, what are some scenarios—and I'm not sure where you can answer this—where poor risk management could lead to being unprepared against terrorist acts?

Mr. YIM. I think that there are several scenarios that we have. When we talk about, for example, the bioterrorism attack in an urban area, poor risk management leads to everyone trying to do the same things. So we can have a lot of different entities begin to stockpile chemical, biological protective suits as the military did following the gulf war in 1991. And we can come back 7 years later in 1998 in the military and find that the shelf life had expired in most of those suits and our protection was illusory.

I think if we don't have coordinated activities, we may have the illusion of greater preparedness, but not the actual reality of being able to fulfill the responsibilities. I think that is an example, sir, that is very troubling for us. People need to enhance capabilities over a long period of time, not just the capability to do something within a particular budget cycle.

Mr. RUPPERSBERGER. Any other examples?

Mr. YIM. I think cyberterrorism—that if we have systems that have identified security holes, and that we don't have coordination so that there can be cascading impacts, or that the vulnerabilities are not clearly made known because people wish to hold back that information for whatever reason, their share value, etc., that we could have significant impacts from a lack of coordination.

Mr. RUPPERSBERGER. Wouldn't you say that a terrorist act has clearly a cascading effect in other population centers, financial markets, infrastructure? Are we prepared, do you think, based on that scenario at the local and State level?

Mr. YIM. I think we are not to the extent that we should be because we have not completed in general the vulnerability assessments that are required. There are some vulnerability assessments that are being done. It is 2 years after September 11 and 5 years after many of the commissions have recommended or identified terrorism as a major threat. We really need to expedite these vulnerability assessments.

Mr. RUPPERSBERGER. Another question: Which agencies in particular bear the greatest risk with the fewest resources?

Mr. YIM. We have just seen the 2005 submission. We've seen which ones got plussed up and which ones did not a bit. Certainly the Department of Homeland Security is getting, in certain areas, increased funding and some downgrades in others.

I think the Department has been under tremendous stress with this reorganization. I know when we have talked to members of the other Departments, they are not fully staffed in many ways. They are having difficulties responding to some of the deadlines that are self-imposed as well as externally imposed. And they're talking about, well, we just don't have all of the management structure or resources in place. I think that's an area that needs to be looked at.

Mr. RUPPERSBERGER. Based on the President's budget that was submitted yesterday, what areas in homeland security do you feel were cut that would have a negative impact on our security?

Mr. YIM. Just with this, I know there has been a great concern about the way that money is going to be distributed to emergency and first responders, State and local. There have been a lot of debates about trying to have that on a risk management basis as opposed to purely a per capita or fair share type based on population approach. I think that is an issue that bears a lot of watching. Are we funding enough for preparedness at the State and local and private sector, and is it going to the areas that, based on good intelligence data, deserve to receive?

Mr. RUPPERSBERGER. Based on the cuts in the first responders, what negative impact do you feel that would have on national security?

Mr. YIM. Certainly any event is going to be local. Everyone uses that phrase. I believe that wholeheartedly. We need to have our State, local and private sector to be prepared for a wide variety of hazards. If we're not, I think the whole Nation suffers.

Mr. RUPPERSBERGER. If you would be able to recommend to the President to reconsider that cut, what would your main argument be?

Mr. YIM. Again, I don't have enough details on the specifics and the justification, but certainly when we talk to State, local and private sector, that's the people that Congress and the administration serves.

Mr. RUPPERSBERGER. Training, equipment.

Mr. YIM. Training, equipment. It's not only that, but generally being prepared to deal with a wide range of events. And the same type of preparedness in the Midwest for a tornado is going to help us in an explosive attack. The same level of preparedness in California for an earthquake is going to help us on a wide range of attacks.

Mr. RUPPERSBERGER. I represent Maryland's Second Congressional District, city of Baltimore, port of Baltimore, BWI Airport, the tunnels, but one of the—and we did a survey on who had received moneys from Homeland Security, and we checked with every volunteer, career, police departments, all the different areas, and this was about maybe 8 months ago, and I believe the results of the survey was over 73 percent of all those entities had not received a penny from Homeland Security.

But more importantly as it relates to your comment, it seems to me that the No. 1 issue that I personally received from this survey from the police, fire, paramedics was the inability to communicate with different systems of communication. And that is very important in a crisis. And New York City, as an example, the Pentagon and different agencies come together. Have you had the occasion to look at that, and what is your opinion as far as the underfunding of that topic?

Mr. YIM. Our infrastructure technology team has looked at that and the issue of bandwidth and what frequencies and compatibility of the communication equipment the people have studied. The World Trade Center identified areas in which the first responders, because of incompatible equipment, did not know—people inside the towers did not know what the people outside knew and therefore did not receive some of the warnings. That will continue to be a problem.

Are we confident? I think our IT team is fairly confident that technology will be able to solve that problem. What they are concerned about and what we are concerned about is once we are able to talk to each other, what are people going to say to each other; what information are they sharing; what activities are they going to be using to coordinate once they can physically talk to each other.

So the immediate problem, yes, enable communications, interoperability absolutely; but we have to address what are they going to say when they can talk to each other.

Mr. SHAYS. This is our second hearing.

Mr. RUPPERSBERGER. We have had a lot of witnesses before this committee, but I think he's very direct in answering the questions.

Mr. SCHROCK. He sure is.

Mr. SHAYS. I agree with the gentleman. This is our second hearing now on strategies. We had a hearing on standards. And Mr. Ruppersberger as well as others in this committee, we put our name on legislation, and it's now part of the draft of the Select Committee on Homeland Security saying to DHS that they've got to establish standards and get them set up sooner.

So, for instance, in my State, I had asked local communities what had they gotten from the Department of Homeland Security, and they said, no, until I did a little more investigation. And what happens, the Department of Homeland Security had given a substantial sum to the State, and the State had given every department—they had viewed before they set up standards that everyone, fire, police, first selectmen, mayors, all needed better radio equipment. Then protective gear got out to a lot of the communities. And they were getting it from the States, and they didn't realize it was a pass-through.

Our point, and I think it's your point as well, and this is what I was going to ask you, sir, so it's a nice lead-in, you also specialize in the whole issue of standards besides strategies. What is or should be the relationship between national standards and national strategies?

Mr. YIM. I think there is a key relationship, and I wish to be sure when we talk about standards—many people talk about individual product or equipment standards. That is an important aspect; an example, the thickness of a Kevlar vest. You were talking about systems standards.

Mr. SHAYS. Why should Westport, CT, get the same as Stamford, CT, or why should a small town in Connecticut like Canaan up north be getting anything necessarily?

Mr. YIM. That is the goal of national management or systems standards. What you want to identify is that everybody should not be doing the same thing. They should be doing slightly different things, and we have to be developing capabilities. They may be resident in different entities or different locations, but there has to be some way to mobilize that capability together in a time of crisis or contingency.

That is the beauty of national standards. National standards based on the ISO, the International Standards Organization, or the American ANSI standards, they were designed in the manufacturing business, in all honesty, and I think that is a great analogy for Homeland Security. When Ford Motor Co.—in looking at its business model, they had to rely on a whole bunch of people in the chain of command. They had to adhere to certain standards as to the quality of the material and the tensile strength. The part suppliers, they had to adhere to certain standards so when it was incorporated into your Ford, that it operated as expected efficiently. They could rely upon that.

That same standards approach could link the Feds, State and local together. We assign responsibilities to the private sector. We give them performance measurements and self-certification to standards, can you meet them on a consistent, reliable basis so we can depend upon them coming to the table when the Feds need them or the States need them. I think that's the key.

Mr. SHAYS. Is it realistic to expect these strategies that we have been talking about to yield to an overarching concept like containment to guide the long-term effort against terrorism?

Mr. YIM. I think that goes back to the classic feeling that where you sit determines what is the most important thing to you. Where you sit on certain areas, containment may be the most important. Where you sit in other areas, response and recovery may be the most important. I don't think that right that we have some sort of overarching goal, because I think different parts of our sectors, government and society have different priorities, and the strategies have to be flexible enough to recognize those differences.

Mr. SHAYS. I don't know if I agree with you, but then again I don't have any basis to disagree. It would seem to me that you're going to want—I want to ultimately have a sense—are you saying this? Are you saying the good old days of the cold war don't allow us to have a fairly concise sense of strategy? I mean, I worked for a year on what we should do to help cities, and we had pages and pages, and it just got bigger, and then it came down in the end to one thing: We needed to bring businesses in to create jobs and pay taxes.

What I'm asking is, after we develop all these strategies, are we going to find some kernels that are going to be found in each one of these strategies that will be—that is what I'm asking.

Mr. YIM. I would not disagree with that. I think there are some overarching drivers that should be there.

Mr. TIERNEY. I wanted to say something. I'm shocked to think that you would think a Governor is passing out Federal money without letting people know. What Governors are we talking about?

Mr. Yim, this month, the Assistant Secretary For Infrastructure and Protection at the Department of Homeland Security stated that the comprehensive terrorist threat and vulnerability assessment is unlikely to be completed in the next 5 years. What can we do to speed that up, and ought we not be focusing on trying to get something before 5 years are up?

Mr. YIM. I think that most experts would agree that we hopefully will have the luxury of 5 years, but it's unlikely we'll have the luxury of 5 years. Definitely, how do we enforce greater timelines? We obviously don't want a bad product, but it doesn't have to be all or nothing. There are many phases that could be put in to a vulnerability assessment and break it down into manageable chunks, set some milestones. If 5 years is the end state, maybe we ought to live with that, but that doesn't mean nothing can be done or measured within that 5-year period of time. So even if we had a 5-year goal, what is the 6-month goal, what is the 1-year goal, what is deliverable in 2 years, and how often are you going to refresh that?

I think that is the real focus, not on the ultimate end state, the timeframe for the ultimate end state. We may never have an ultimate end state. In 5 years, that is an eon in Washington, DC. It is an eon for most agencies and certainly beyond the life of a political appointee's life. In light of that, I think you have to set interim steps, and that's one of the deficiencies we pointed out.

Mr. TIERNEY. On the homeland security issues, do you see any importance to educating the local populace with respect to reaction to an event? Do you see that as part of the strategy in homeland security? And where do you put that in the level of importance with other things we might do?

Mr. YIM. I think the communication strategy and the education strategy is vitally important. Our citizens have to have confidence in the ability of our governments to protect themselves. If you look back at the anthrax attacks immediately following in October 2001 and the somewhat confusing information that was promulgated, I think that needlessly alarmed or caused people to take actions that perhaps were not only unnecessary, but may have been counterproductive. The broadband use of—or the widespread use of a broadband antibiotic could have other deleterious effects.

And information—I think to the education, what people need to do for their own protection how they interact with other people, things that they can't do, there's no way that I'm going to be able to protect against a nuclear attack or even be able to really protect myself against a nuclear attack, and I'm just going to have to live with that, but other things I could do.

Mr. TIERNEY. Did you see much of that in the Homeland Security Strategy?

Mr. YIM. It doesn't get down to that level of granularity. I think the National Security Strategy talks about an education component. I don't recall that any of the other strategies deal with an education component, and I think that's a gap.

Mr. TIERNEY. Thank you, Mr. Chairman.

Mr. SHAYS. Mr. Schrock.

Mr. SCHROCK. Mr. Yim, the chairman said that—he talked about Stamford, CT, and talked about Canaan, CT, and I thought I heard you say maybe everybody doesn't need to be doing the same thing, which indicated to me that you thought maybe the small towns didn't need to worry as much as some of the bigger areas. But over the holidays, while the big areas were getting a lot of chatter, the Town of Tappahannock, VA—it is a wonderful little town, very, very small town, and there was a lot of chatter that was going to be a target. Now if something had happened there, what does that say for every little berg and town in America? It is out in the beautiful boondocks, I should say, but it's a magnificent place. But what does that say for other little towns that might think, oh, my gosh, now we are going to be targets, because towns in the Midwest think they are safe, and they are not.

Mr. YIM. I think there are some minimal levels of preparedness that no matter where you are, we would expect our towns to be able to respond in certain matters, and because of resource constraints or their location, maybe the only thing we are asking them to do is do a holding action until other resources can be mobilized and arrive. I think that's part of the strategy development. You can't expect that small town to defeat a major bioterrorism attack, but maybe they can triage the patients and hold them in isolation for 48 hours until something can arrive.

Mr. SCHROCK. Mr. Ruppersberger talked about the ports. Port security is my No. 1 issue. I represent the Port of Hampton Roads, which is the Norfolk, Virginia Beach area, and every ship that comes into the massive port has to pass by the largest naval base in the world. And I worry about somebody trying to sneak a major container ship behind our piers and lock our ships in, sort of like what they did in Pearl Harbor.

But the good news is the port of Hampton Roads has been the guinea pig, and I think a very good guinea pig, for all this new equipment to test all the containers and the trains that come in and out of there with some absolutely incredible results, and it is the only port in America that has it right now. And the test results are so good, I can see it going to other ports as well. Frank and I went down there a few weeks ago, and I was amazed at the progress.

The bad news is they have to pass by the Navy before they get there. The good news: They are being screened if they think something is wrong. The port of embarkation is where everything needs to be.

Thank you, Mr. Chairman.

Mr. SHAYS. Mr. Murphy, are you all done?

Thank you very much. You have been a wonderful witness, and I'm assuming you or someone from your staff will be able to hear the panelists.

Mr. YIM. I will stay myself.

Mr. SHAYS. We are going to invite you to come back and make some comments, so don't fall asleep during that second panel. Thank you very much.

Our second panel is comprised of four individuals: Dr. Lani Kass, professor of military strategy and operations, National War College; David H. McIntyre, former dean of faculty, National Defense University; Colonel Randall J. Larsen, U.S. Air Force, retired, CEO of Homeland Security Associates; Mr. Frank Cilluffo, associate vice president for homeland security, the George Washington University.

You just stay standing. We will swear you in.

[Witnesses sworn.]

Mr. SHAYS. Our witnesses have responded in the affirmative, I'm sorry you are kind of crunched in. We are going to start as you are on the table. Dr. Kass, we will go with you first, and we will do the 5 minutes. We would like you to stay somewhat within the 5 minutes, but we could go over to the next 5. We prefer it closer to 5. You have the floor, Dr. Kass.

STATEMENTS OF LANI KASS, PROFESSOR OF MILITARY STRATEGY AND OPERATIONS, NATIONAL WAR COLLEGE; DAVID H. McINTYRE, FORMER DEAN OF FACULTY, NATIONAL DEFENSE UNIVERSITY; RANDALL, J. LARSEN, COLONEL, USAF (RET), CEO, HOMELAND SECURITY ASSOCIATES; AND FRANK CILLUFFO, ASSOCIATE VICE PRESIDENT FOR HOMELAND SECURITY, THE GEORGE WASHINGTON UNIVERSITY

Dr. KASS. Thank you, Mr. Chairman, gentlemen. I'm an American by choice rather than the fortune of birth. I'm particularly honored to be here. And I'm going to avail myself of your generosity of a little bit more time because otherwise I will require English subtitles.

The views I'm about to present are my own. They reflect over 20 years experience as a teacher and practitioner of strategy. They do not necessarily reflect official positions of the U.S. Government.

Let me start with a quick historic vignette, if I could. In May 1863, on the eve of the Battle of Chancellorsville, General Joe Hooker, Commander of the Union Army of the Potomac, said, "My plans are perfect. May God have mercy on General Lee, for I shall have none." General Hooker's overconfidence had immediate mid and long-term consequences. First, he was crushed by General Lee. Second, he was fired by President Lincoln. Third, General Hooker did not go down in history as a great strategist. Instead his name became a synonym for, shall we say, certain ladies of the evening.

Mr. SHAYS. You didn't tell me you were going to be entertaining.

Dr. KASS. The joint lesson, Mr. Chairman, is that humility is a virtue when assessing strategic plans, your own or anybody else's. That is so because, simply put, strategy is hard to do. Strategy seeks to balance ways and means. It seeks to mitigate risk. It seeks to account for current imperatives, future contingencies and unpredictable dynamics of human behavior. Thus it operates in a realm where chance and fog and friction and ambiguity dominate. Everything in war is very simple, but the simplest things are difficult.

Strategy guides action. It needs to be translatable into a series of implementing plans, but it cannot be so specific as to delve into tactics. It is supposed to provide vision. It is supposed to provide what in the military is called commander's intent.

Strategic effectiveness—and, Mr. Chairman, you asked about that—comes from a synchronized effort sustained over the long term and guided by a clear vision of what it is you are trying to accomplish, what is called the desired end state; in other words, how do you want the situation to look when you are done doing what it is that you are doing.

Foresight and flexibility are the keys to success. So is the ability to integrate a wide variety of variables into a coherent whole. In short, Mr. Chairman, this kind of holistic thinking is pretty uncommon primarily because it is so difficult. A logical systematic approach is the necessary first step.

I provided the committee with what we use at the National War College to educate the Nation's future strategic leaders. Hopefully it will help you and your staff ask the difficult questions that need to be asked when evaluating any strategic design.

The first strategic question and the most comprehensive is to assess, to understand the nature of the war you're engaging in. What then is the nature, the character of the war we are engaged in? And I will focus the rest of my remarks on this.

Clearly terrorism is not new. It has been with us for a very long time. What is new is that modern technology has provided individuals with destructive power which up until now was the sole domain of advanced militaries. What is also new is that choice can now operate on the global scale in pursuit of global objectives. With the world as their battleground and globalization as their enabler, they seek to destroy the American way of life and the international system we lead; that, Mr. Chairman, what we are fighting is an insurgency of global proportions, what I would term a pansurgency. This insurgency is not tied to geographic boundaries. Instead it operates in nontraditional domains using nontraditional means clearly and bound by accepted norms of civilized behavior. This insurgency has invoked a legion to declare war on the United States and to mobilize the sympathies of 1.5 billion Muslims.

The breathtaking scope of the insurgent goals is matched by their desire to inflict casualties virtually anywhere on the planet. They seek weapons of mass destruction and would not hesitate to use them. They are willing to destroy everything and die trying. They're well financed, exquisitely networked, adaptive, flexible and patient. They also know us much better than we know them.

The ultimate defeat of this global insurgency will only come from the synchronized application of all instruments of national power guided by an overarching strategic design and not a practical plan. We must defeat terrorist organizations which have global reach. We must deny them sanctuary and State support. We must diminish the conditions that allow terrorism to flourish. And we must do all that while defending the homeland. So what we're talking about is a multidimensional strategy which fuses offensive and defensive and integrates all elements of national power.

Mr. Chairman, I truly believe that terrorism is the societal evil of our time. The war on terrorism is our generation's greatest challenge. This evil must be abolished just like slavery, like piracy, like genocide. We are engaged in a war which demands the long-term commitment of the Nation's will, blood and treasure. It also demands a consistent, focused strategy to achieve the end state of abolishment of terrorism. That does not mean every individual act. Slavery was abolished a long time ago, and there is still slavery, piracy and genocide in the world. But that is the end state we should strive to. And the mission of any current strategy is to provide you this overarching end state that you are trying to achieve.

The American people and your elected representatives should not expect a quick or easy victory. I believe World War II and the cold war are pretty useful to think about in terms of the scope and magnitude and duration of the fight we are engaged in. The war on terrorism is a war of necessity which we must win.

Mr. SHAYS. Thank you very much.

[NOTE.—The National War College report entitled, "Combating Terrorism in a Globalized World," may be found in subcommittee files.]

[The prepared statement of Dr. Kass follows:]

Statement of

Dr Lani Kass

Professor of Military Strategy

National War College, National Defense University

February 3, 2004

House of Representatives

Committee on Government Reform

Subcommittee on National Security, Emerging Threats, and International Relations

Mr. Chairman, thank you for inviting me to testify. As an American by choice rather than the fortune of birth, I'm particularly honored to be here. The views I'm about to present are my own. These views reflect over 20 years of experience as a teacher and practitioner of strategy. They do not represent the official positions of the US Government, the Department of Defense, or the National Defense University.

In May 1863, on the eve of the battle of Chancellorsville, General Joe Hooker, commander of the Union Army of the Potomac, said: *"My plans are perfect. May God have mercy on General Lee, for I will have none."* General Hooker's over-confidence had immediate, mid- and long-term consequences: First, he was crushed by General Lee. Second, he was fired by President Lincoln. Last, General Hooker did not go down in history as a great strategist; rather, his name became a synonym for--shall we say--certain ladies of the evening. The enduring lesson is that humility is a virtue in assessing strategic plans--your own or anybody else's.

This is so because strategy is hard to do. Strategy operates in a realm where chance, fog, friction, and ambiguity dominate. It seeks to reconcile ends, ways and means; mitigate risks; and balance present imperatives with future considerations--all in an uncertain, dynamic environment. To complicate life even further, strategy is a multi-sided affair. This means that the objectives, intentions, and reactions of both allies and opponents are difficult--if not impossible--to anticipate and account for. Clausewitz was right: "Everything in war is very simple, but the simplest thing is exceedingly difficult." We should remember this principle as we evaluate current strategies and look for ways to do better in the future.

Strategy guides action. Thus, it is nothing but pragmatic. The focus of strategy is on how to use available means to achieve the desired ends with acceptable risk. Therefore, the first strategic question is: will this idea--this "perfect plan"--work under the special--and, usually, unknowable--circumstances of its next test. Often, that next test is the crucible of war.

Innovation, flexibility and integration are the hallmarks of successful strategies. The ability to think anew and develop creative solutions to changed circumstances is as critical as it is rare. Innovation hinges on foresight--that is, the ability to assess current and emerging trends, as well as anticipate their potential. Innovation requires courage, perseverance, and, often, readiness to "break some china"--especially in large bureaucracies.

Integration--or, "holistic thinking"--is an approach which captures both the whole and its component parts; grasps multi-dimensional, dynamic relationships as they are today and as they might evolve tomorrow; yet does not assume--nor expect--perfect coordination, clear-cut answers, or immediate, measurable results.

Moreover, successful strategies must be linked both upward and downward. The best military operation will be an abject failure if it does not support the over-arching political strategy. Likewise, a brilliant strategy unsupported--or unsupportable--by reality at the tactical and operational levels is, at best, an interesting academic exercise or, more often, a prescription for disaster.

Strategy is both an art and a structured intellectual process. Strategic effectiveness comes from an integrated, synchronized effort, sustained over the long-term, and guided by a clear vision of the desired end-state--of what it is that you are trying to achieve. Foresight and flexibility are the keys to success, as is the ability to fuse a wide variety of actions, issues, and equities into a coherent whole. Frankly, this kind of holistic thinking is rare, precisely because it is difficult.

A logical, systematic approach is a necessary first step. I have provided the Committee with the framework we use at the National War College to educate the Nation's future leaders. Hopefully, it will help you and your Staff ask the tough questions that must be answered to validate the suitability and feasibility of any strategic design.

Asking the right questions is vitally important precisely because the Global War on Terrorism is a new kind of war imposed on us by a new kind of an enemy. This enemy is not tied to geographic boundaries; instead, it operates in non-traditional domains, employing non-traditional means, clearly unbound by established norms of international behavior. This enemy invoked religion to declare war on America--indeed, on civilization at large.

Clausewitz teaches that "the first, the supreme, the most important act of judgment that the statesman and the commander have to make is determine the kind of war upon which they are embarking, neither mistaking it for nor trying to make it into something that is alien to its nature. This is the first strategic question and the most comprehensive." What, then, is the nature of this new war?

Clearly, terrorism is not a new phenomenon. What is new is that modern technology provides individuals with destructive power that up till now was the exclusive domain of advanced militaries. What is also new is that terrorists can now operate on a global scale, in pursuit of global aims. With the world as their battlefield, and globalization as their enabler, these insurgents want to destroy the existing international system and establish a new world order, dominated by their brand of militant Islam. Thus, we're faced with a new strategic equation: an insurgency of global proportions--what I'd call a PANSURGENCY--meaning a networked, transnational **movement**, aimed at overthrowing values, cultures, and societies by means of terrorism, subversion, and armed conflict.

The breathtaking scope of the insurgents' goals is mirrored by their desire to inflict mass casualties, virtually anywhere in the world--be it New York or Riyadh; Washington or Nairobi; Dar-al-Salaam or Jerusalem; Bali or Baghdad; Ankara or Jakarta. They seek weapons of mass destruction and will not hesitate to use them. They truly believe they're on a mission from God; they are ready to destroy everything and die trying. They are well-financed, networked, adaptive, flexible, and patient. They also know us much better than we know them.

What is also new is the explicitly religious nature of al Qa'eda's ideology. Religiously motivated violence is different for the simple reason that, for the true believer, there is no compromise about the sacred; there can be no bargaining, nor accommodation, nor truce. In this context, killing becomes an end in itself, rather than one instrument among others, to be used rationally to attain the desired objectives. Thus, this first war of the 21st Century is as deeply rooted in the ancient past as it is in the imperatives of the information age. It might also be the first deliberate effort to re-introduce religion into international relations since the 1648 Treaty of Westphalia effectively banished considerations of creed from the repertoire of acceptable reasons to wage war.

It has been said that terrorists want a lot of people watching rather than a lot of people dead. If so, religiously-motivated terrorists are fundamentally different: They do want a lot of people dead and may not care whether a lot of people are watching, as long as God sees what has been done in his name. God's partisans cannot bargain over the fulfillment of his will, because doing so would substitute man's judgment for God's. In this construct, total, global war is simply unavoidable.

The last time the US fought a "hot" war on a global scale was 60 years ago. Like the current Global War on Terrorism, the Second World War started with a surprise attack on US territory; it called for a total commitment and quick adjustment to unexpected imperatives. In both cases, US forces faced conditions they had not planned on or prepared for, requiring us to adapt in the midst of a fight, learn from experience, and quickly evolve new approaches and procedures--and, often, field new, untested technologies--to solve emerging problems. Intellectual agility and strategic adaptability--the ability to innovate--along with war-fighting and organizational skills, proved to be the keys to victory in WWII. These very same skills will be necessary to win the Global War on Terrorism.

The Second World War and the Cold War are useful paradigms to think about the Global War on Terrorism--in terms of scope, duration, the desired end-state, and, most importantly, the level of national will and commitment that were required over the long haul. It is also important to remember that both World War II and the Cold War were battles of ideas: democracy and capitalism against fascism and communism. In the end, our ideas triumphed: fascism and communism were relegated to the "ash heap of history." That is where al Qa'eda's ideology belongs.

Terrorism is the societal evil of our time; the war on terrorism is our generation's greatest challenge. **This evil must be abolished and universally delegitimized like slavery, piracy, and genocide**. There must be an international taboo against the deliberate targeting of innocent civilians. We must create a global environment hostile to both terrorist organizations and terrorism.. Though acts of terror can never be completely prevented, terrorism must be reduced to a level that is isolated, rare, and clearly irrational--that is, useless as an instrument of policy. This will ultimately allow terrorism to be combated as criminal activity within single states, not as a global war.

The defeat of this global insurgency will only come through the synergistic, steadfast, and systematic application of all the elements of national power--diplomatic, economic, informational, financial, law enforcement, intelligence, and military--simultaneously across four dimensions: We must ***defeat*** terrorist organizations; we must ***deny*** them sponsorship, support, and sanctuary; we must win the battle of ideas and ***diminish*** the underlying conditions that allow terrorism to flourish--all while ***defending*** the US.

The centers of gravity of terrorist groups include their leadership, supporting ideology, finances, command and control network, and sanctuaries. To defeat them, the United States, its allies, and coalition partners need to: Identify and isolate terrorist organizations at each level; Disrupt their support infrastructure and sanctuaries; Discredit theirs ideology; and Destroy their networks and leadership.

While it is unrealistic to hope to eliminate every single terrorist who desires to threaten innocent civilians, it is possible to eliminate the synergy created by cooperation of disparate terrorist organizations. This effort will reduce the operational scope and capabilities of global and regional terrorists to the point that they become threats only at the individual state level. At this level, the threat can be combated as criminal behavior, which will allow a narrower focus to attack their centers of gravity and allow full engagement of law enforcement mechanisms.

The second element of the Strategy of Abolishment focuses on deterring future acts of terrorism. To establish a credible deterrent, the United States and the international community should develop and maintain capabilities and mechanisms that clearly communicate to potential terrorists and their supporters that the costs of action would far outweigh any perceived benefits. The deterrent message should be sent not only to terrorist organizations but also to states that sponsor them, to nonstate actors that provide a front for their activities, and to individuals who may contemplate joining or supporting them. Deterrence and denial support the strategic aim of abolishment by convincing individuals, organizations, and states to seek alternate methods of political change because terrorism is no longer a viable option. Sending an effective message to each of the four audiences associated with terrorism requires:

Deterring terrorist organizations. Terrorist organizations believe that they can conduct operations with impunity. Capabilities, particularly improved intelligence, should be acquired to detect, thwart, and destroy such groups and bring their members to justice. Actions should be taken to create certainty that terrorists will be captured and imprisoned rather than becoming "martyrs" for their cause. Political, social, and religious leaders must understand that their organizations will be destroyed if they choose terrorism to advance their aims.

Deterring state actors. Terrorist organizations must be denied state support or sanctuary. This can be accomplished by demonstrating the resolve to replace the leadership of any state that continues to sponsor terrorism, as well as by broadening international norms against terrorism.

Deterring nonstate actors. Nonstate actors must be deterred from providing aid and assistance to terrorist organizations. This can be achieved by establishing an international environment of greater financial transparency, "naming and shaming" organizations involved in terrorist support, and lowering barriers to asset seizures and freezing of funds.

Deterring individuals. Efforts to deter individuals from joining or supporting terrorist organizations include educating potential recruits on the sinister nature of specific organizations and of terrorism in general, dispelling the notion that terrorism results in any positive gain, and demonstrating that terrorists will be brought to justice.

Although some believe that terrorists are undeterrable, a strong argument can be made to the contrary. Without question, state and nonstate actors can be deterred from providing assistance. The tougher challenge applies to the actual terrorist organizations and their followers. Ultimately, however, terrorists must be compelled to believe that their efforts would be futile--or face certain destruction.

Efforts to diminish the underlying causes of terrorism comprise the third element of the Strategy of Abolishment. Through an intensive, long-term campaign, the United States and its allies should strive to mitigate the underlying conditions that foster the formation of terrorist groups and allow them to recruit their support elements. To do this, the United States and its allies should engage vulnerable regions and disparate ideologies and peoples.

The major contributors to the underlying causes of terrorism are: Economic and social inequities in societies marked by both abject poverty and conspicuous affluence; Poor governance and economic stagnation or decline that alienates many segments of a state's population; Illiteracy and lack of education that lead to widespread ignorance about the modern world and resentment toward Western values in general and US foreign policies in particular.

To mitigate these conditions, the US must engage in and win the war of ideas. A sustained information campaign should denigrate the concept of terrorism and discredit its supporting ideology. Concurrently, we should increase foreign development assistance and use it to promote accountable and participatory governance, along with sustained economic growth, literacy and education in the Islamic world and underdeveloped nations.

While the United States engages in overseas activities to combat terrorism, it should simultaneously defend the homeland. The US faces an enemy determined to disrupt the American way of life and undermine the safety and security of US citizens everywhere. On the home front, the United States should remain vigilant and ready by establishing collaborative relationships between Federal agencies, law enforcement, public health and emergency management entities, professional associations, and private partners. To that end, the United States should use every measure available to defend the homeland against terrorist attack, while executing its overarching offensive strategy to abolish terrorism. The US should be postured to provide an effective defense in three areas:

Prevent terrorist attacks. To the maximum extent possible, would-be terrorists and the weapons they intend to use must be denied entry into the United States. Weapons of mass destruction must be detected and intercepted before they can be employed. Collaboration at all levels of government, along with private sector and individual citizens, is essential to disrupting terrorist aims.

Protect critical assets. To minimize the probability of a successful terrorist strike in the homeland, the United States should fortify critical infrastructure and other potential terrorist targets.

Prepare responses. To reduce the impact of terrorism, the United States should be prepared to mitigate the consequences of an attack. This is particularly critical in respect to WMD attacks. Again, collaboration among all agencies at the Federal, state, and local level is essential.

Integration of the offensive and defensive aspects of the strategy is key. The US should be safe and secure at home to preserve its way of life, maintain economic growth, and remain engaged in the international counter-terrorism effort. Without an effective defense, the United States might be driven to focus on matters at home, allowing terrorists to continue operating on a global scale.

In sum, this Nation is engaged in a war that demands a long-term commitment of national will, blood and treasure. It also demands a well-orchestrated, consistent and focused strategy to achieve the desired end state: A world free of organized terrorism and a global environment in which terror can never again flourish. The American people and their elected representatives should not expect a quick or easy victory. Yet, we must all realize that this is truly an existential fight--a war of necessity, which we must win. Simply put, failure is not an option.

NATURE OF WAR

- We are truly in an existential fight
- This is not about what we do; it's about who we are.
- WW II and Cold War are useful parallels in terms of both duration and end state.
- Long term fight to abolish terrorism
- WE ARE NOT AT WAR WITH ISLAM. BUT RADICAL ELEMENTS IN MUSLIM WORLD ARE AT WAR WITH US.

WHAT TERRORISM IS?

- Politically motivated violence deliberately targeted at civilians.
- Instrument of social and political change.
- Terror seeks to break people's will, so they surrender principle to save themselves.

WHAT TERRORISM IS NOT

- Terrorism is NOT senseless or random.
- One man's terrorist is NOT another's freedom fighter.
- Terrorism is NOT a viable negotiating technique.
- Terrorism doesn't seek compromise.

MODERN TERRORISM

- Is networked, adoptive, decentralized, ruthless and lethal.
- Current technology gives <u>individuals</u> the capability to wreak havoc on a scale hitherto only large, advanced militaries possessed.
  - They seek WMD and will use them.
  - Our only viable strategy is "no quarter"
  - Carry the war to the enemy

WHY THEY HATE US?

- They don’t hate us for what we do.
- They hate us for who we are and what we stand for—the beacon of democracy, freedom and modernity.
- They want to restore the caliphate.
- They want to restore Islamic glory by the sword.

THE TERRORISTS' ADVANTAGES

- Are ready to destroy everything—and die trying
- Have sympathy—if not support—of 20% of world's population
- Are hard to find—some live among us
- Are well financed
- Are patient
- Know us much better than we know them
- Really believe they're on a mission from God.

WHY WE'LL WIN?

- We can't afford to lose.
- This is a war of necessity—what's at stake is everything we are.
- We're the future, they're the past
  - They exclude 50% of their own society
  - The reward they offer is death
- Others always underestimate our will and resolve:
  - In matters of national survival, we're determined and violent—just ask the Germans and Japanese.

END STATE

- Terrorism ABOLISHED and DELEGITIMIZED like other evils
  - Slavery
  - Piracy
  - Fascism/genocide
- International taboo: killing of innocents is NEVER a legitimate means of political change.

INTEGRATED STRATEGY

- Need to break fundamental asymmetry wherein we need to succeed 100% of the time and they need to be successful only once.
- Best defense is good offense.
- Don't start developing strategy from point of failure
  - Seize the initiative—shape the war

INTERCONNECTIVITY

TERRORIST CATEGORIES

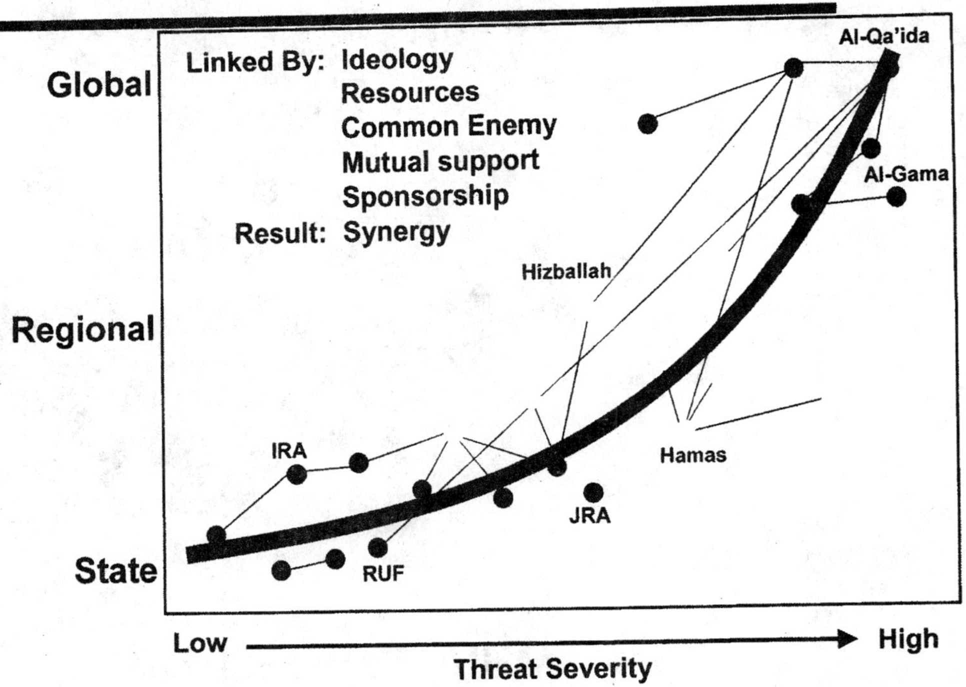

GLOBAL VIEW

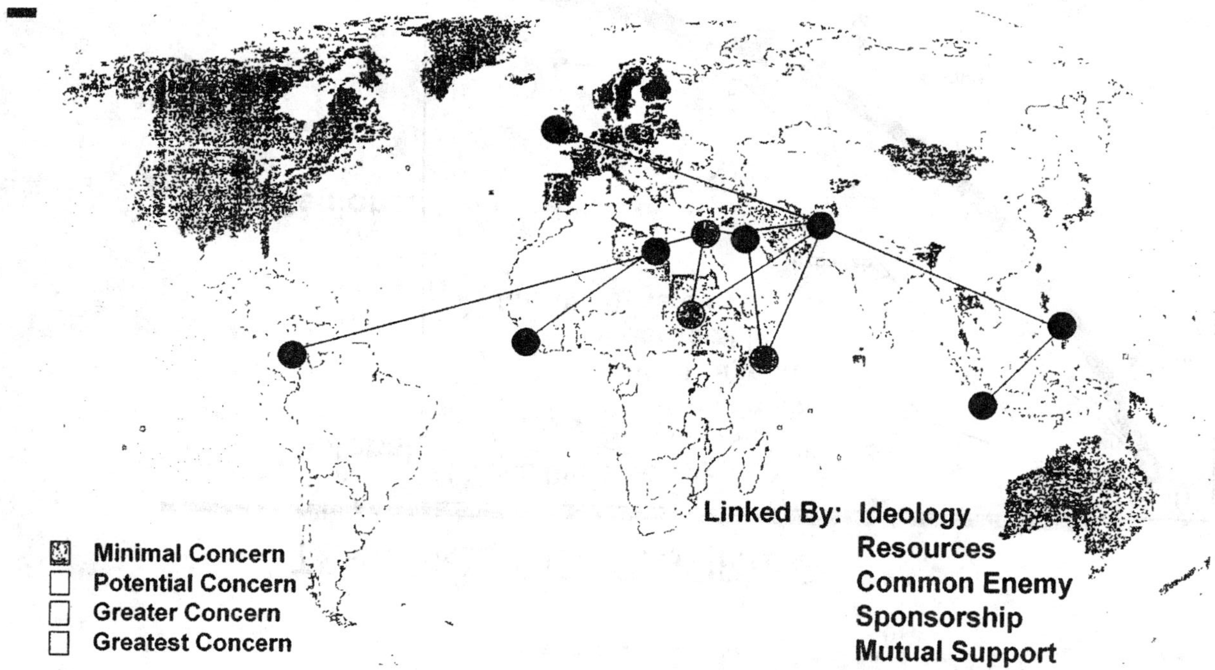
Minimal Concern
Potential Concern
Greater Concern
Greatest Concern
Linked By: Ideology
Resources
Common Enemy
Sponsorship
Mutual Support

A NEW KIND OF WAR

OLD Paradigm: **Terrorism, guerrilla warfare, conventional warfare used as tools to foster change *within* a society or state - Insurgency**

NEW Prism: **Terrorism, guerrilla warfare, conventional warfare used as tools to foster change *across* societies and states - Pansurgency**

"Pansurgency" Defined

Organized movement of *transnational actors* seeking to overthrow values, cultures, or societies on a global level through subversion and armed conflict, with an ultimate goal of *establishing a new world order*.

CONCEPT OF "PANSURGENCY"

Insurgency of Global Proportions

- Overthrow existing societies
- Inter-linked terrorist networks
- Few nations untouched

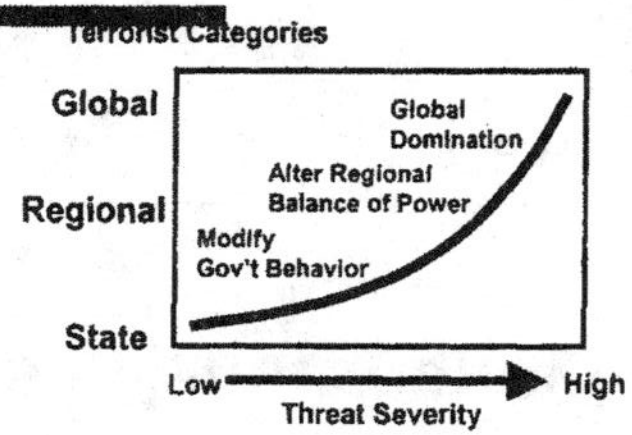

"Globalized" Nature

- Virtual Nation-state with instruments of power
- Strategy transcends international boundaries
- Jeopardizes security of peaceful societies
- Legitimacy derived from ideology
- Attempts to rally 1.5B Muslims against Western culture

PANSURGENCY THREAT

ENDSTATE

- Overthrow existing societies
- Global domination

OBJECTIVES

- Unite Islamic World against Western culture
- Obtain WMDs
- Defeat Israel
- Remove" infidels" from Islamic lands
- Incite worldwide insurgencies

CENTERS OF GRAVITY

- Leadership
- Finances
- Ideology / "Cause"
- Safe havens
- Command/Control
- Linkages

"3 D" CONCEPTUAL FRAMEWORK

Defeat existing terrorist groups by:

- Disrupting their support infrastructure
- Discrediting their ideology
- Destroying their networks and leadership

Deter terrorist groups by:

- Detecting potential acts of terrorism
- Denying their ability to execute those plans
- Defending against those actions should deterrence fail

Diminish underlying causes of terrorism by:

- Determining contributing factors to global terrorism
- Directing actions to mitigate those factors
- Decreasing potential for "blowback"

DEFEAT EXISTING TERRORISTS

<u>Goal</u>: Eliminate state sponsorship of terrorism and reduce organized terrorism to a level manageable as crime within the boundaries of a single state.

- Identify and isolate terrorist organizations at each level
- Disrupt their support infrastructure
- Discredit their ideology or rationale for committing terrorism
- Destroy their networks and leadership

DETER FUTURE TERRORISTS

Goal: Maintain the required set of capabilities that serve to convince potential terrorists that risks far outweigh any perceived benefits of engaging in acts of terrorism.

- Implement regimes/protocols to dissuade terrorism
- Detect potential acts of terrorism through integrated systems
- Deny terrorists the ability to carry out plans
- Provide effective defense should deterrence fail

DIMINISH THE CAUSES

Goal: Reduce underlying causes and remove expectation that terrorism will result in political, ideological, or material gain.

- Determine and mitigate factors that give rise to terrorism
- Engage in information operations to denigrate terrorism
- Establish mechanisms to ensure terrorism achieves no gain
- Decrease potential for "blowback"

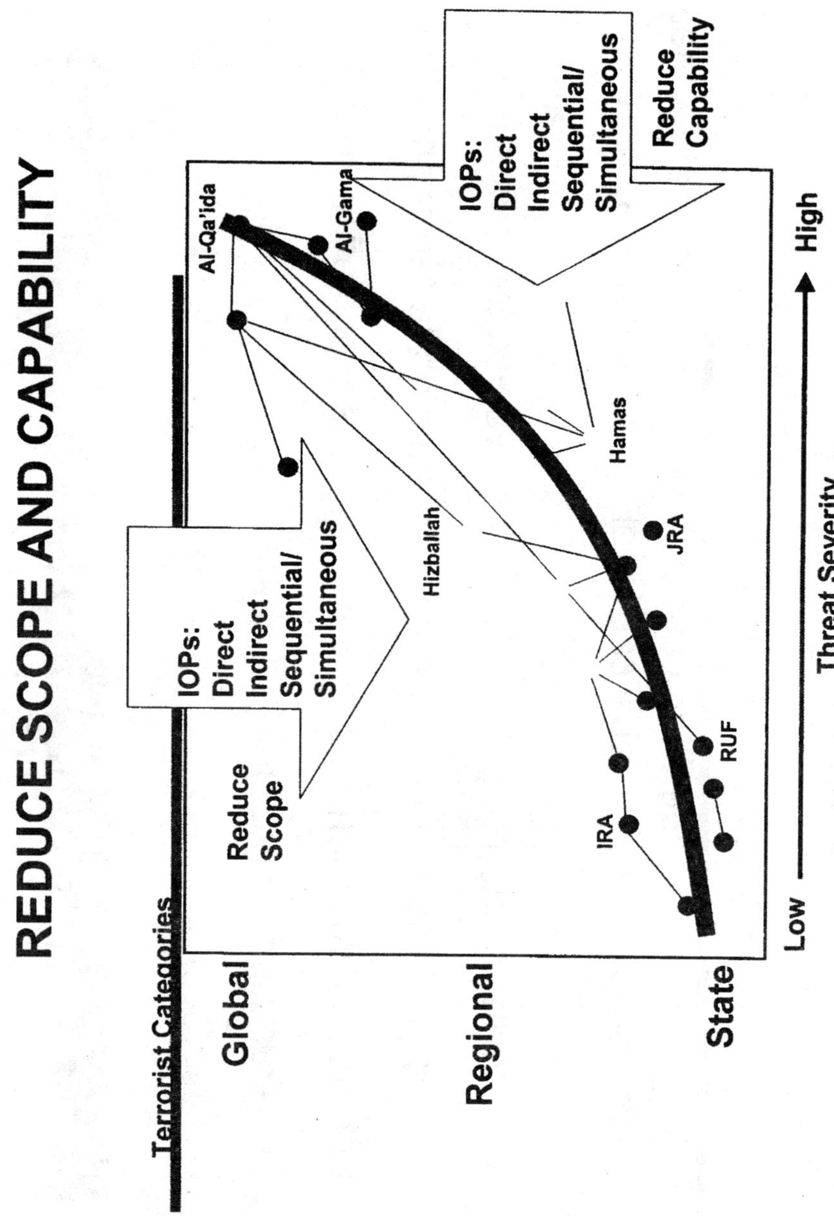
REDUCE SCOPE AND CAPABILITY
Terrorist Categories
Global
Regional
State
IOPs:
Direct
Indirect
Sequential/
Simultaneous
Reduce
Scope
IOPs:
Direct
Indirect
Sequential/
Simultaneous
Reduce
Capability
Al-Qa'ida
Al-Gama
Hamas
Hizballah
JRA
RUF
IRA
Low
High
Threat Severity

REDUCE SCOPE AND CAPABILITY

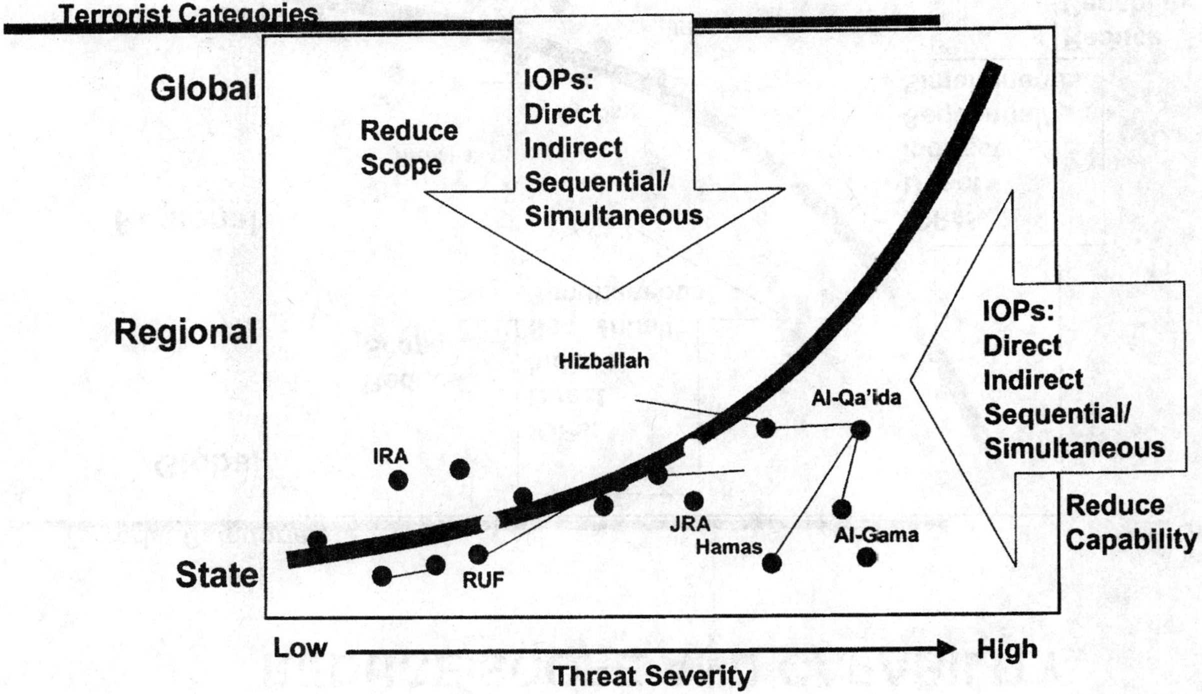

REDUCE SCOPE AND CAPABILITY

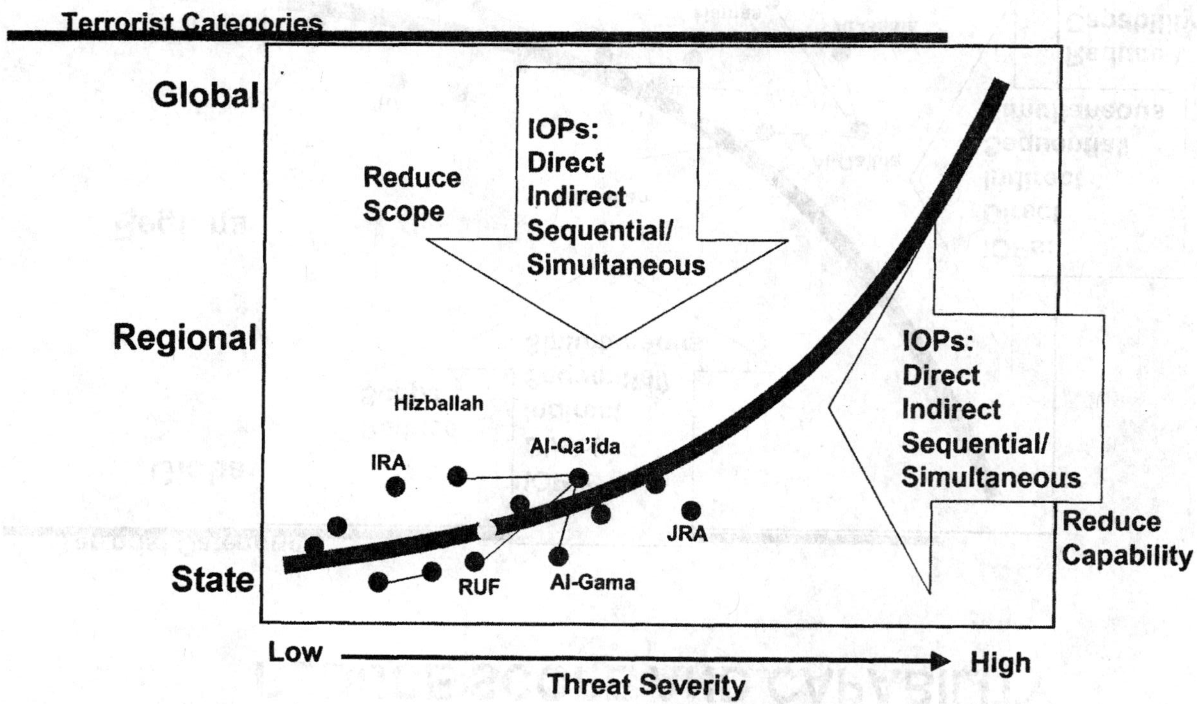

STRATEGY ENDSTATE

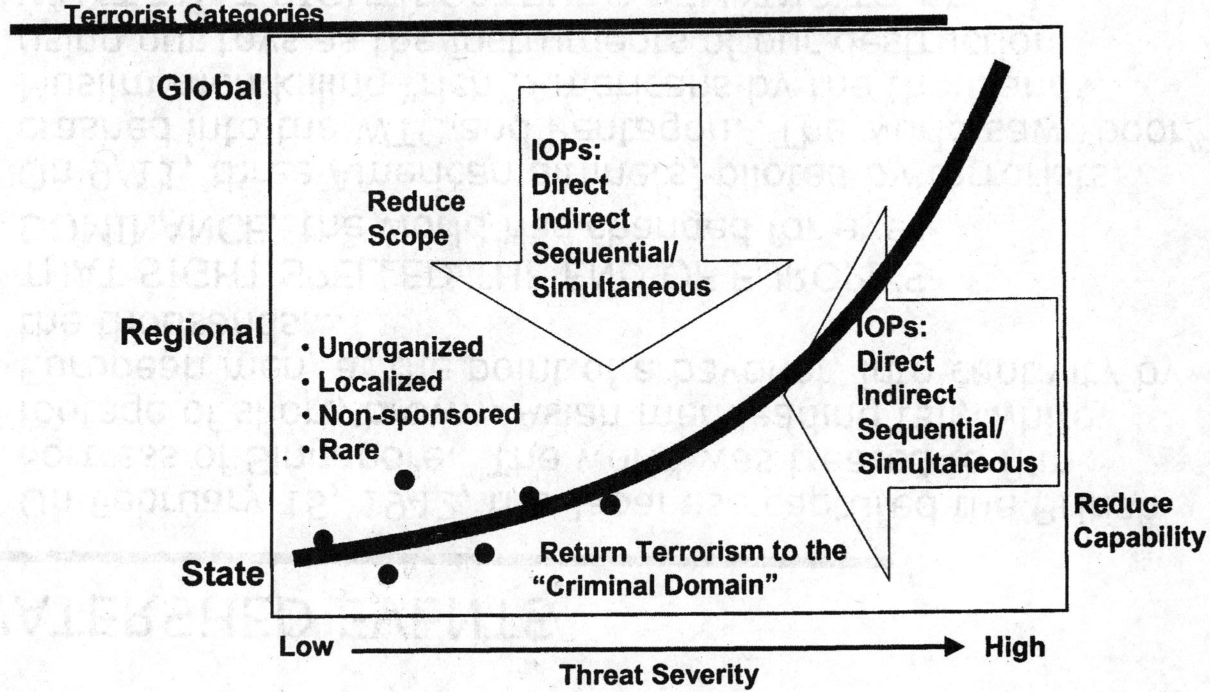

WATERSHED EVENTS

- On February 15, 1942, the Japanese captured the British Fortress of Singapore. The world was treated to film footage of short, brown, Asian men leading tall, white European men, at the point of a bayonet, into captivity by the thousands....
- THAT SIGHT SPELLED THE END OF EUROPE'S DOMINANCE...the world has changed for ever.
- On 9/11, three American airliners, piloted by terrorists, crashed into the WTC and Pentagon. The world saw "poor" Muslim men killing "rich" Americans by the thousands, using our toys as the instruments of our destruction.
- WHAT THAT SIGHT PORTENDS REMAINS TO BE SEEN....BUT THE WORLD WILL NEVER BE THE SAME.

Mr. SHAYS. I have to say you are the most honest witness I have ever had, because usually when I say you have 5 minutes and please don't roll over another 5, everyone says, well, I will try to stay within the 5 minutes. And you just said, I'm not even going to try to stay within the 5 minutes, so I think you got away with it because of your unique accent.

Dr. McIntyre.

Dr. MCINTYRE. I will try to stay within the 5 minutes even though my accent is from Texas.

Mr. Chairman and members of the committee, I want to thank you for the opportunity to testify today on the important subject of our strategies for national security and homeland security in the war against terrorism, and I want to thank you in particular for the work this committee has done on this subject in the past.

The United States is involved in a new, long war for its survival, but it does not feel like war, so voters don't always give full credit to the elected officials and appointed officials who wage it and oversee it. I suspect the members of this committee do not always get the credit due them for the effort spent on these subjects so critical to the long-term destiny of our Nation. As a former military officer and current student of strategy who spent long years studying what happened to nations contemptuous of strategic realities, let me thank you for your efforts in this field.

In my written statement I have tried to do what you asked, use my 36 years of strategic and military experience to conduct an analysis of the family of strategies prepared by this administration, evaluate their adequacies both individually and collectively, so I will only summarize.

I think the administration's approach to offering a family of strategies to formally lay out their goals in many areas and their concepts for achieving those goals is an admirable one. I recognized what they're doing immediately. It looks like every major military plan I have ever seen. I am not uncomfortable with what some have called the proliferation of strategies.

I will give you five brief points we will have to address as time proceeds and we look to refine our strategies. No. 1, we must clarify the fact that in the short run this is about managing dangers to America and attacks upon Americans, not eliminating them. In a world where technology gives big weapons to small people, we cannot eliminate every threat. Some attackers will get through. Some innocent people will become casualties. This is not failure, this is reality. We need to prepare the American people for this reality.

No. 2, clarifying the forcing function that will eventually reduce or eliminate terrorist attacks on America is key. This is a tough one because it involves changing the nature of the enemy. We have to cause him to lose hope of victory through what he's doing and accept some alternative solution to his grievances. This is not even easy to conceptualize, certainly not easy to do, but this is the essence of the long-term victory.

No. 3, because we cannot kill every potential enemy and protect every potential target, we must prioritize our spending and our efforts. I recommend that our highest priority go to preventing and responding to the types of high consequences of attacks that will be the most damaging to the Nation as a whole. Without a set of public priorities, we will be drawn constantly forward to expanded actions overseas and expanded spending at home. The biggest problem in this war will be knowing where to stop. We need to set these priorities in public.

No. 4, we must give more attention to the enemy. Many people and even some experts are still operating under the cold war assumption that our enemies' grievances have to do with economics and the distribution of wealth. That is fighting the last war. This war is about ideology and legitimacy. In the long run, we are going to have to offer an alternative to the enemy's ideology. I am not confident that we have yet considered the implications of that fact.

And finally, we must understand that this war will be waged over generations. We cannot win it if we change our underlying strategies every time we change administrations. During the cold war, we pursued a strategy of containment for 40 years through a variety of administrations. The actions, the priorities, the expenditures changed from administration to administration, but not the underlying strategic concept that by denying communism growth and additional resources we would doom it.

As in the cold war, we need strategies that will stand the test of time. They must be bipartisan strategies that can garner support across party and ideological lines, and that is why the work of this committee is so important. Thank you again for your efforts in this regard and for the opportunity to contribute to that effort.

Mr. SHAYS. Thank you very much.

[The prepared statement of Dr. McIntyre follows:]

Statement of
Dr. David H. McIntyre (COL, USA, Ret)
February 3, 2004
House of Representatives
Committee on Government Reform
Subcommittee on National Security, Emerging Threats, and International Relations

"Strategies for a New Long War: Analysis and Evaluation"

In the mid-1990's, the Chairman of the Joint Chiefs of Staff, General John Shalikashvili, testified before Congress that the United States had reached a point of "strategic pause" in its relations with the rest of the world. No clear enemy existed with both the capability and the intent to strike US vulnerabilities overseas, much less at home. Consequently, he argued, the focus of US military thought and acquisition should be the type of force we would want to field in the year 2010. Later refinements pushed out the focus of technological research and doctrinal development to the year 2020. Every military staff and most military colleges devoted themselves to identifying "the next big thing" – using information technology to reshape the military to face an unknown "peer competitor" twenty years away, with a strategy that called for "domination" of any foe in any form of combat. If only we could dominate the battlefield with precision fires, maneuver, intelligence and logistics, the logic went, the enemy would be deterred from attack, and destroyed in short order if a fight were required.

At almost the same moment that the Department of Defense declared the near term horizon free of threats, Osama Bin Laden was meeting with his chief operatives to lay out ideas for an attack on the US homeland, and a new long war designed to collapse the American economy, will, and civilization. Americans might have been reassured by the failure of attackers to down the Twin Towers in New York in 1993, but Bin Laden was emboldened. While the Americans overlooked the developing threat and sought to build a freer and more prosperous world by enlarging free markets and democracy, Bin Laden and others worked feverously on a strategy to destroy moderate Muslim regimes, fracture the community of civilized nations, and collapse the "infidels" who supported modern Islamic leaders.

It was a strategy that almost worked. It might yet.

The Bush Administration has responded to this new strategic situation with a variety of short and mid-term programs, from military action to destroy terrorist sanctuaries in Afghanistan, to diplomatic and law enforcement action to choke off the funding of terrorist training and operations worldwide. And they have published a family of national strategies – a set of nested concept papers, that lay out lines of thought as well as specific actions to address the new strategic situation.

This is a new approach to crafting and presenting policies for the future, and some, more comfortable with the narrow challenges of the past, profess themselves confused by the "proliferation of strategies." But I for one am pleased to see a set of public plans laid out for review with the intent of coordinating our government, intimidating our enemies, and informing our citizens. That does not mean I agree on every point – but I do applaud the boldness of stating the ideas in a coherent manner and opening the field to analysis of plans and results.

McIntyre
03 February 2004

In this paper, I will conduct such an analysis and offer a judgment on the strategies developed thus far, based on my thirty years of military experience, and sixteen years of crafting, studying, and teaching strategy at the national level.

What Is Strategy and How Does It Work?

We will begin with a brief review of the subject of national strategy, only because the term is so often abused and the fundamental ideas so often misstated. Most frequently, strategy is called a plan to balance ends and means.[1] But this definition, I conclude, is wrong – or at least incomplete.

1) The first component of a good strategy is a **clear concept of where the leader wants to go** – what end is to be achieved. Only in a well-defined war between well-defined enemies does a national level strategy have an endpoint. In times of peace (and quasi-war) the desired "end" is usually the management of a problem, not its solution

2) A good strategy is based on a concept of **cause and effect**: *IF I want X to occur, **THEN** I must do Y to make it happen.*

This seems a simple point, but it is at odds with most strategic teaching and practice today. Even experts and senior officials become so engaged in "operationalizing" the strategy (i.e., crafting policies and carrying them out), that they often forget this first critical piece. A successful strategy must be built around a forcing function – some concept that will cause the stated goal to be achieved. And if the goal is to keep a problem manageable rather than pay the price to solve it, then that should be stated up front.

3) Once this fundamental concept is in place, the **actions and resources** to achieve it – the **ways and means** – must be allocated. Here is where most of the action lies on a day to day basis. Tactics, operations, logistics, personnel training, education and management, organizational and doctrinal development, and coordination with others (intra-agency, interagency, interjurisdictional, and international), prioritization and budgeting – all lie in this part of the strategy. Action oriented leaders are naturally attracted to this process, and many outside observers (especially the media) look to this area alone to evaluate effectiveness. But without a good concept of cause and effect as a base, policy making can become disassociated from logic. The result is action, but not strategic action -- process without progress

4) A good strategy must **allow for a thinking enemy**. It must focus on success, not just action. It must reduce enemy capability and will, as well as reduce friendly vulnerability and strengthen our capability and resolve. This requires some system for measurement, periodic review and adjustment.

5) And finally, **strategy takes place over time**. This is more a question of establishing perspective than setting a timeline. The Cold War took 50 years. It might have taken a decade less, or two decades more – there was no way to anticipate the timeline. But the logic of our strategy did appear compelling, and included patience and the passage of time as critical elements of its success from the beginning.

---

[1] Joint Pub 1-02, DOD Dictionary of Military and Associated Terms, The Joint Staff, Washington DC., defines terrorism as: "The Calculated use of unlawful violence or threat of unlawful violence to inculcate fear; intended to coerce or to intimidate governments or societies in the pursuit of goals that are generally political, religious, or ideological."

To summarize, we will use the following structure to evaluate national strategies in the remainder of this paper:

Framework for Analysis of Strategies
1) Does the strategy establish a clear end?
2) Does the strategy establish a clear and compelling cause and effect as a forcing function.
3) Are appropriate programs and resources provided for implementation?
4) Is the enemy considered? Is there a way established to periodically review whether we are being strengthened and the enemy weakened? Is the strategy periodically adjusted as a result?
5) And finally, is the strategy designed to work over time?

A Case Study in National Security Strategy: Containment

The intellectual framework we need to analyze today's family of strategies is better understood by looking at a well known, widely accepted, and wildly successful example: the strategy of "containment" with which we won the Cold War.

As with our situation today, the situation in which national strategists found themselves in 1950 was entirely different from what they expected.

At the end of World War II, many senior Americans expected the UN to prevent future war, the United Kingdom to patrol the world as before, and the US to return to a comfortable role as a partner focused on economic advantage.[2] But US Ambassador to the Soviet Union George Kennan shattered this comfortable view with his famous "long telegram" from Moscow, and a later article in Foreign Affairs by "Mr. X," in which he described the emerging hostile global ideology that put the survival of the US at risk, together with a potential response. Events along the "Iron Curtain," in Berlin, in China, and finally in Korea, convinced skeptics that strong action was required. Following the North Korean attack in 1950, staffers at the National Security Council, led by Paul Nitze, codified Kennan's ideas into a strategy that became known as "Containment."

The core of containment was not a balancing of ends and means, but a concept of cause and effect. Communism was a fundamentally flawed idea, the strategy argued – because it misread the nature of man, it could only redistribute wealth and power, it could not create them. So ***IF** the US could cut off Communist nations from new resources and populations, **THEN** the whole communist edifice would eventually collapse of its own internal contradictions.*

The actual employment of the strategy suffered from a variety of interpretations from the very start. Nietz favored a robust approach to containment, while Kennan favored a more benign form of diplomacy. The argument over how containment should be operationalized ricocheted through government offices, think tanks, and academia for years, and not just between liberals and conservatives or Democrats and Republicans. In the Reagan administration, Secretary of Defense Weinberger and Secretary of State Schultz played out differences not unlike those of Nitze and Keenan. And in the process, over a 40 year period when the fundamental strategic construct of engagement was set, but the way it was to be realized was endlessly debated, this

[2] Paul Nitze and Nelson Drew, NSC-68: Forging the Strategy of Containment, National Defense University, 1994.

debate – the debate over how the concept of cause and effect was to be operationalized – the argument over what ways and means were to be employed to achieve the ends envisioned – came to be regarded as the making of national security strategy itself.

More money to defense or to education? More carriers or peace corps workers? Conventional troops or nuclear weapons? Once the Containment Strategy was codified in 1950 by NSC-68, these questions over purchases and priorities constituted the whole of the strategic argument for the whole of the Cold War. And so arguments over balancing means and ends (or ways, means and ends as military strategists prefer to say), are taught today as the stuff of fundamental strategic analysis.

And to be sure, a whole family of strategic decisions did follow from this fundamental concept of strategic cause and effect. For example:

- Once the decision was made to fight the hostile ideology around its whole periphery, the decision followed to size the military and the government for the fight. Beginning in 1950, the US built a conventional military force large enough to surround the Communist world, and prepared to fight, either conventionally or with nuclear weapons if necessary. Deterrence, forward deployment, military alliances, and Mutually Assured Destruction were all part of a military strategy to support the strategy of Containment worldwide, as was the whole process of raising, training, equipping, educating and employing an enormous federal bureaucracy outside the military, ranging from the Department of State, to intelligence agencies, to a robust industrial base. The entire economy of the nation was involved..
- The strategy also included a decision to pay for the new standing military. In 1947, President Truman sought to reduce the defense budget to $7 billion. Three years later it was seven times that size, and it continued to grow in keeping with the need to build a global force.
- Because of the need to field a large force and pay for it, and because Communism was as much a moral challenge as a physical one, NSC-68 included provisions to mobilize the will and resources of the American people.

To summarize:

Strategy Evaluation: Containment
1) Does the strategy establish a clear end? Containment did establish a clear end – the end of communism -- the destruction of the hostile ideology
2) Does the strategy establish a clear and compelling cause and effect as a forcing function? Yes, it did establish a clear and compelling cause and effect as the forcing function.
3) Are appropriate programs and resources provided for implementation? Resource levels and support varied over time, but adequate funding for the strategy was agreed to in principle for the entire 50 years of its existence.
4) Is the enemy considered? Is there a way established to periodically review whether we are being strengthened and the enemy weakened? Is the strategy periodically adjusted as a result? The strategy proved very elastic regarding enemy measures and counter-measurers – actual execution was adjusted frequently, without hurting the coherence of the strategy itself.
5) And finally, is the strategy designed to work over time? The designers of the strategy expected it to work over time. Communist leaders were dangerous but not suicidal. So we could afford to take our time – to win without preemptive action or precipitating a war.

To be sure, the strategy was adjusted many times over the next five decades. But its key decisions and structures were put in place early on. And it was possible early on to identify the underlying concepts as adequate, even if the employment of that strategy varied in quality over the years.

For Comparison: A Brief Evaluation of the Clinton Strategy

When President Bill Clinton took office in the first heady days after the end of the Cold War, he identified three primary threats to the nation:

- The proliferation of weapons of mass destruction
- The resurgence of old totalitarianism in newly democratic countries
- Excess military spending that robbed the nation of resources required for other priorities.

After an 18 month delay, his administration produced a national strategy of Engagement and Enlargement. The underlying assumption was that rich democracies do not fight each other. *So the national strategic concept was that **IF** the US would use its national level resources (to include its military, its position at the UN, etc) to promote justice, freedom and prosperity around the world, **THEN** America would be safer and more prosperous as a whole.*

The National Military strategy was designed to compliment this strategic concept with an approach called "Shape, Prepare, and Respond:"

- Using military forces in particular to engage other nations and shape their development toward democratic ideals;
- Preparing military forces for the future by saving money now (keeping the size small and holding down acquisitions), while planning for a "Revolution of Military Affairs"(RMA) that would produce a military force both cheaper and more effective in the long run;
- And maintaining adequate forces to respond to crises as they emerged.

Beginning in November of 1996, The Clinton administration added a refinement that both clarified the national strategy, and significantly expanded its scope. After implying as much for four years, the administration explicitly identified securing and expanding US national values as a matter of US survival. This raised the stakes for every US interaction overseas, and placed the Department of Defense and others on a virtual wartime footing in support of every aspect of engagement.

As it turned out, the Clinton strategy was a bit naïve about the power of economic and political incentives to change political opportunists and deeply rooted hatreds. Additionally, the requirement to defend our values everywhere all the time as a survival issue made prioritizing very difficult. The resulting burden of global engagement on deployed troops was greater than anticipated, even as the RMA turned out to be more expensive than expected – requiring even further limits on manpower and stretching our modernization programs to stay within budget.

Additionally, in a move not unlike the current development of multiple subordinate strategies, the Clinton administration identified a number of emerging security needs at the federal level, and addressed them with a variety of Presidential Decision Directives (PDDs);

- PDD 18 and 37 laid out an approach to counter proliferation of Weapons of Mass Destruction.
- PDD 39 looked at the challenge of Transnational Threats (whether crime, drug trafficking, or the threat of terrorism).

- PDD 56 established a set of interagency committies to address different domestic crises ("Complex Contingencies") in order to practice for crises in peace, and promote rapid cooperation in emergencies.
- PDD 62 and 63 established the intellectual and organizational framework to designate and (eventually) promote the protection of facilities identified as "Critical Infrastructure".
- And PDD-63 established new organizations and initial thoughts about how to promote public-private partnerships in pursuit of improved cybersecurity for the nation.

These PDDs began the organizational efforts of what came to be called federal homeland security, but the task identified was massive, and the resources devoted to these new duties were never adequate.

In short, the combination of stated strategy and directed organizations turned out to be both more problematical and more expensive than anticipated. And the public strategy was blind to a series of threat developments (specifically the threat of Islamic terrorism) that really required some entirely new strategic concepts, ways and means. The international aspect of the strategy was well grounded in theory, and appreciated by many other nations who saw it as cooperative in nature. But the resources were inadequate to the task. So despite some notable successes in the short term, the strategy could not succeed in the long run without considerable new expenditures – most of which were carefully scheduled to come due after President Clinton left office.

To summarize:

Strategy Evaluation: Engagement & Enlargement
1) Does the family of strategies establish a clear end? President Clinton's Engagement strategy was really a way to reorder the world power structure, not a way to engage or defeat a particular threat.
2) Does the strategy establish a clear and compelling cause and effect as a forcing function? Engagement did establish a clear concept of cause and effect, but that concept was theoretical, not proven. In fact, engagement caused some resentment in some places, and the "democratic peace theory" upon which it was based is now in question.
3) Are appropriate programs and resources provided for implementation? Resource requirements turned out to be considerably greater than anticipated. The strain on the military was particularly noticeable.
4) Is the enemy considered? Is there a way established to periodically review whether we are being strengthened and the enemy weakened? Is the strategy periodically adjusted as a result? The strategy was designed specifically to overcome the resistance of opponents through a variety of regimes and multilateral actions. The ability to tailor approaches by nation was a strong point of the strategy. And the Clinton administration should receive credit for recognizing that potential domestic security challenges would require a different, more interagency response. But identifying issues to US values as survival challenges made reforming the world a life-and-death issue, and prioritization nearly impossible.
5) And finally, is the strategy designed to work over time? The concept of a time line did not exactly apply to this strategy, since engagement was seen as an end unto itself. The strategy did consider time in that it was intended to last until a sufficient number of nations accepted free markets and democracy to make those concepts the norm around the world. But the project was essentially open ended, continuing until the very nature of the international system was reformed.

In the final analysis, this last point colors the overall analysis of the strategy. If the goal was to keep nations engaged and talking rather than fighting, then perhaps "Engagement" can be considered a success. But the language of the strategy seemed to promise a global revolution of major proportions, so the strategy raised more expectations than it could deliver. In the process, it expended a considerable part of the intellectual capital and physical resources of the military chasing marginal improvements in the US security posture around the world. And beyond that, the chances of profitable engagement with the dangerous fanatics at that moment conspiring to attack the US in its homeland were non-existent. The strategy that attempted to shape a new world in a "moment of strategic pause," ultimately proved inadequate in a world already being shaped by the twisted logic of fanaticism.

An Overview of the New Family of Strategies

As previously noted, I find the family of strategies issued by the Bush administration to be a big step forward in public accountability. Certainly, many of the government's plans and strategies remain secret as they should. But this approach of nesting strategies gives an excellent view of what the administration considers important, what it is willing to do to achieve those important goals, and what it is not. In this section we will conduct an overview of most of the strategies, and then examine whether and how they work together to advance US national interests.[3 4]

The National Security Strategy of the United States of America

The Bush administration's strategic approach began in an entirely new context:

- A new technical revolution gives big weapons to small people.
- A new global revolution means big challenges not subject to traditional solutions.
- A new terrorist revolution means big new enemies with a small footprint.
- A new ideological conflict stakes a claim to one fifth of the world's population and poses a major danger to survival in the long haul.

Developed in the aftermath of 9-11, the strategy must both advance US interests in the world, and address a survival threat to the nation and Western Civilization.

So this administration has developed an entirely new approach to deal with this situation. The goal appears to be not destroying an enemy (as in the Cold War), or reforming the world (as with the Clinton administration), but managing the threat.

The major goals identified for overseas are not new, but the focus is: the new emphasis is to take actions by others into account in shaping our interaction with them. This approach is much more accommodating to those who cooperate with us than those who oppose us. And the focus of our assistance is not primarily on the most needy nations (as in the past), but on those most likely to reform. The strategy is not intended to reshape the international system, but to advance and secure America's position in that system.

[3] This paper does not consider the 2002 National Money Laundering Strategy, as such an evaluation requires a special level of financial expertise.

[4] This paper does not consider the National Military Strategic Plan for the War on Terrorism because the level of enemy action and the certain existence of classified plans in augmentation must surely be causing modification on a daily basis. For example, just 5 days before this testimony the Secretary of Defense changed his long standing policy against expanding ground forces and allowed expansion of the Army by 30,000 troops over the next 4 years. Discussing these wide changes in policy orally can be very profitable. Analyzing them in writing when only part of the fact are known is more problematical.

This is not a coldly selfish strategy. In fact, the new strategy professes to benefit all who are friendly to free markets, democracy, and the rule of law. But while the Clinton strategy sought to benefit the world and secure the US in the process, the Bush national strategy seeks to benefit the US, producing a more peaceful and prosperous world in the process.

Evidence of this focus would include:

- The strong support for free markets in every nation.
- A strong emphasis on new investment policies opening markets to outsiders.
- A requirement that those who desire the advantage of cooperation with the US develop transparent financial systems so investments can be tracked.
- And a new emphasis on the rule of law as a prerequisite for US engagement, not a product of it.

The national security strategy clearly considers the threat of "terrorism with a global reach," but is not driven by this consideration alone – the goal is a strong, secure, prosperous and competitive America. But the clear recognition that modern technology can be used by a new type of vicious enemy requires a new approach to security: "proactive counter proliferation"

The fundamental arguement is that given the new catastrophic threats abroad (biological war, covert use of nuclear weapons, etc.), we cannot delay action until a clear threat turns into an attack. Logic demands that we be ready to preempt if we have good intelligence and are confident that the danger is real.

Although the willingness to consider preemption has garnered great attention, this is not the core of the new strategy. Only in exceptional cases is preemption anticipated. But the acknowledgement of such potential cases is a major break with the past, and the administration is sensitive to charges that it is acting as a "rogue nation" based on its strength and not international law. The solution is to expand the accepted international doctrine of "imminent threat" to justify preemption in special cases. This careful distinction has not mollified critics.

In particular, it is important to understand that preemption is <u>not</u> "the new strategy." Preemption is merely part of a larger, more traditional strategy that seeks to expand the US circle of friends overseas. But two items really are new:

- Giving first priority for US assistance to those making successful efforts to help themselves.
- Expressing publicly a readiness to act unilaterally and preemptively to meet major threats to the US, to include regime change among selected enemies if appropriate

So the new US National Security Strategy, really looks like this: ***IF** we put US interests first, focusing on areas and issues where those interests are most endangered, working with others where possible but independently if necessary, **THEN** enemies will decline and friends will increase, and both the US and the world as a whole will benefit.*

The National Strategy for Combating Terrorism

Nested within the new national security strategy is a fundamental decision: engage in a Global War on Terrorists with Global Reach – but do so by attacking the terrorists and their support physically, without either mobilizing the American people, or engaging the hostile ideology.

The strategy barely mentions Islam or the radical theology which underlies the motivation of our most dangerous enemies, attempting instead to make war on their actions. It does so with a layered program of actions:

- <u>Defeat Terrorists & their Organization</u> specifies that we will use force to Attack, Destroy, Degrade, Disorganize, Disperse the enemy. This is an extremely proactive use of force to kill terrorists and keep them on the run
- <u>Deny Sponsorship, Support, Sanctuary</u> suggests that these proactive measures will be pursued whether host nations like it or not.
  - Those willing and able to defend themselves will be helped.
  - Those governments willing but weak will receive support.
  - Those reluctant to cooperate will be "convinced".
  - Those unwilling to cooperate will be coerced.
- <u>Diminish the Underlying Conditions</u> calls for an international effort to assault the political and economic conditions encouraging individuals to embrace an ideology hostile to the US and its interests. Note especially that:
  - The US calls on the international community to assist in this effort.
  - The strategy does not address the underlying religious arguments that prove such a powerful motivator for many of those who have attacked us.
- Finally, <u>Defend US Citizens & Their Interests at Home Abroad</u> makes it clear that the Global War on Terrorism will be waged globally.

Taken together, this is an extremely proactive strategy: ***IF** we Defeat, Deny, Diminish and Defend a wide range of enemies and potential attackers worldwide, **THEN** attackers will be so reduced that we will all be safer.*

The logic holds up in a mechanical, absolute evaluation: reducing and eliminating enemies means fewer enemies in the long run. But the key motivating factor for our most dangerous enemies seems to be religious. The failure to recognize and address this fact, while smoothing relations with moderate believers around the globe, leaves a glaring hole in our strategic logic. It also means that the single most important metric to measure our success over time – the reduction in the scope and impact of radical teachings – will not be considered for evaluation.

The implications of this strategy for resources are significant but not explicit. Perhaps this is to be expected. NSC-68 included in 1950 size and budget estimates for the forces to be employed for the strategy of Containment, but these estimates remained classified for 25 years. Perhaps the Military Strategy for Combating Terrorism, and other classified documents contain similar estimates. And perhaps keeping them classified is a good idea – no reason to let the enemy know what burden he is placing on our economy. But this is an expensive part of the strategy family, and the absence of any public estimate of needs and costs leaves the administration in the position of saying "just trust me" to the Congress and the people. Perhaps a requirement for periodic reports to Congress on resources needs would be advisable.

The National Strategy to Combat Weapons of Mass Destruction

The last administration did work hard to counter the proliferation of WMD, using a wide variety of approaches, from establishing international regimes, to pressuring governments to give up their programs, to paying foreign scientists to do other work. The new National Strategy to Combat WMD continues these approaches, but makes one profound change. It commits the US to direct action to secure or destroy WMD that pose a direct threat against the US. In the President's words, "We will not permit the world's most dangerous regimes and terrorists to threaten us with the world's most destructive weapons." Key issues include: Interdiction, Deterrence, Mitigation, and Defense (both proactive and preemptive)

Of course, Interdiction and Deterrence are not new in this game, but *the explicit threat to hold all who work on such programs personally responsible is new*. So is the emphasis on proactive defense (to include missile defenses), and preemptive defense (again an extension of established international law concerning "imminent threat"). There is no question that under this strategy, the administration is prepared with operational capabilities to neutralize threats overseas should negotiations fail.

These bold warnings about overseas action are matched by a list of responsibilities for consequence management at home, and the assignment of the Secretary of Homeland Security to direct and coordinate Federal efforts. The actual list of actions is a bit short and mundane for practical application, but the significance of including preemptive action and homeland security as elements of a strategy to counter WMD is great. Clearly the administration is taking this threat very seriously, and staking out a position of resolve – no one should be surprised at subsequent preemptive action.

In this regard, publishing the strategy and putting potential WMD proliferators on notice may well be part of the strategy itself.

The National Strategy for Homeland Security

This is really a plan for action, not a strategy. Its emphasis is on organization, responsibility, accountability, and preventing unintended consequences – all ways and means rather than ends. The underlying construct is for managing the problem of terrorism, not constructing a logic of cause and effect to eliminate it. And the principle tool of management is to be the Department of Homeland Security (DHS).

As the subsequent Strategy for Critical Infrastructure Protection makes clear, many federal agencies beside DHS will have key roles in preventing and responding to terrorist attacks in the US. But by virtue of the 22 agencies selected for consolidation, the new department establishes a set of priorities at the national level. These include: Intelligence and Warning; Border and Transport Security; Critical Infrastructure Protection; Response to Catastrophic Threats; Emergency Preparedness & Response; and Domestic Counterterrorism. Not surprisingly, this list corresponds closely to the organization of the new department, which itself foreshadows the core of the administration's budget request for homeland security. The result is a sort of "strategy by organization," where the critical cause-and-effect relationship that defines the administration's strategy for homeland security may be distilled from its actions and priorities. ***IF*** *the new agencies within DHS are properly resourced and accomplish their missions,* ***THEN*** *the survival of the nation will be assured, even if the safety of all individual citizens is not.*

This is a rather convoluted way of discerning exactly how this part of the family of strategies works, and the entire enterprise would be greatly improved if the administration would simply lay out the need to prioritize the security of its citizens, and explain its vision for doing so. But perhaps this is politically untenable – I notice that the administration's critics have not laid out their priorities either.

Three other areas receive special emphasis in the Homeland Security Strategy:

- Federalism: The strategy repeatedly emphasizes the constitutional limitations on what the federal government can direct and control. While individual federal agencies have significant power, the primary exercise of that power will be through establishing standards, grant programs, and incentives for state, local and private cooperation. Command and control of Homeland Security is largely a local issue, and this strategy means to remind and reinforce on this subject.

- Cost: The administration is determined that homeland security not break the federal bank, and it emphasizes that future costs are likely to be shared equally with state and local jurisdictions, and private industry (each entity paying about 1/3 of the anticipated $100 billion annual cost.)
- Accountability: The goal of making every element of the entire system accountable for both its effectiveness and its efficiency is excellent, but easier to promise than to achieve. With the strategy in place now for nearly two years, the number of individuals held publicly accountable for poor performance has been low.
- Restraint: To the administration's credit, the natural tendency of federal agencies to grow themselves and constrain others is recognized, and a specific caveat emplaced that "America and American Freedoms" must remain unchanged. Provisions for specific, periodic reviews in this area would have been useful.

The National Strategy to Secure Cyberspace

This strategy has been the most difficult to craft, and remains the most problematic in the administration's entire family of strategies. The reasons are twofold:

- Offensive tools are advancing more rapidly than defensive tools in the area of cyber security. The fear of an unexpected "Cyber Pearl Harbor" rises daily.
- But the government does not own or control the vast majority of assets at risk, nor can the government secure resources critical to the nation without the cooperation of public and private agents who frequently have little short term incentive to do so.

In short, the federal government can exercise leadership in this area, but success depends upon private action. Crafting a cause-and-effect relationship under such circumstances is nearly impossible. So this document is more a national exhortation than a national strategy.

On the other hand, the strategy makes an excellent effort to organize mission impossible. Protective actions are categorized and addressed in five areas: federal government; state and local government; major industry: small business; private user. Because of the structure of the internet and information revolution. federal directives can only be issued to federal agencies. But the list of recommended actions provides an excellent backbone for action by any organization or individual seeking to secure his own assets and contribute to the cyber security of the nation.

The strategy makes heavy use of "Information Sharing and Analysis Centers" (ISACs) – public-private partnerships encouraged by the federal government but sustained by members of the private sector. ISACs provide a forum where carefully screened professionals can share information with the federal government to improve their security, without conducting meetings that would trigger provide sensitive information to terrorists through our open press.

Other major elements of the strategy include creation of the following:

- National Cyberspace Security Response System: A network of overlapping networks that ties together strategic and tactical analysis of events and trends from the DHS operations center to the network of ISACs, and on to state, local, and private entities that have expressed interest and expertise. This formal network linking informal networks attempts to provide federal coordination for a huge variety of non-federal networks, their plans and operations. It is a massive and frustrating effort, clearly demonstrating the difficulty of securing assets when you can only encourage, and not control
- National Cyberspace Security Threat & Vulnerability Reduction Program: The idea is to create a better process for identifying cyber threats and vulnerabilities, and alerting the public

in general and some participants in particular to the danger. This would collect at a single federal agency the responsibilities and capabilities now exercised by a variety of softwear designers, vendors, service providers, etc. This effort took a major step forward just last week when the cyber security division of DHS announced a new federal alert system that will make the government the trusted source of computer-security information.

The strategy expresses the intent to expand such program internationally, but little progress has been made in this direction thus far.

In short, the strategy recognizes the need for new organizations, plans, resources, and does consider the ever changing nature of enemy attacks – it is written to respond to constantly changing enemy actions for the foreseeable future. But it essentially substitutes information sharing for information control, making a virtue of necessity.

The National Strategy for the Physical Protection of Critical Infrastructure

No strategy provides a better example of the new strategic realities than this one. Given that every locality thinks that its facilities are critical, this strategy provides a major service in identifying the subject as "large scale damage, causalities, damage to national prestige, morale, confidence," prompted by attacks on:

- Eleven critical infrastructure sectors: Agriculture and Food; Water; Public Health; Emergency Services; Defense Industrial Base; Telecommunications; Energy; Transportation; Banking and Finance; Chemicals and Hazardous Materials; Postal and Shipping.
- Five key asset categories: National Monuments and Icons; Nuclear Power Plants; Dams; Government Facilities; Key Commercial Assets.

This definition is at odds with many state, local and private definitions, and even some other federal offices. Merely by issuing this definition, the strategy begins to force a consensus on what receives priority for protection, eventually moving all jurisdictions toward the three objectives it identifies:

- Protection for the most critical infrastructure.
- Protection against high risk specific threats.
- A program of continual evaluation and cooperation at every level.

As with other nested strategies, the principles in CIP are: federal guidance; decentralized execution; information sharing.

Specific actions accomplished by the strategy include:

- Assigning responsibility within the federal government for protecting federally owned infrastructure.
- Assigning lead responsibility in the federal government for coordinating the protection of infrastructure owned by others (while providing as many specifics as possible)
- Providing assistance to owners in the state / local / private sector in their security efforts.

In short, the strategy identifies major issues (thereby giving them priority); and works through ISACs to share information, encourage solutions, and promote "enabling initiatives."

The requirement for resources is not highlighted in this strategy, nor is the enemy. As strategist Colin Gray has observed, "Strategy is so difficult to design and do well that considerations of an intelligent and self-willed foe is frequently a complication too far."[5]

That would appear to be the case with Critical Infrastructure Protection.

---

[5] Gray, Colin, *Modern Strategy*, "Chapter 1: The Dimensions of Strategy," Oxford University Press: NY, 1999, P. 42.

McIntyre
03 February 2004

In Summary

In responding to a survival challenge we have never faced before, the Bush administration has attempted to do something never done before: lay out a family of nested strategies to provide explanation, direction, and continuity to its international and domestic policies. In doing so, the administration opens itself to critics who might take issue with one element or another. Continuing in the face of such criticism – essentially taking a chance on being embarrassed in public –shows a high degree of confidence on the part of the administration, and an admirable determination to get these issues under control.

I have provided an analysis of each strategy individually. I intend to evaluate the effort as a set. I do not undertake this evaluation lightly. Shaping and implementing multiple strategies while waging a Global War On Terror is a bit like changing the tire on a moving car. My hat is off to the leaders and staffers who conceived and recorded them.

Nonetheless, I do take issue on some points.

Strategy Evaluation: Bush Administration Family of Strategies
1) Does the family of strategies establish a clear end? Taken together, the strategies do point to a new end state, where the threat of major terrorist attacks is diminished, and the US continues to dominate the international system. The strategies seek to manage the world, not reform it, and to negate the actions of the enemy, not the ideology that spawned him. On this last point the strategy appears to me to be too narrow and too optimistic. No one on any side of this fight – neither Republican nor Democrat, neither conservative nor liberal - has rushed to grasp the nettle at the center of this conflict: the role of Islamic thought in producing and sustaining the fanatics at war with us. This deficiency must be addressed.
2) Does the family of strategies establish a clear and compelling cause and effect relationship as a forcing function? Yes, but on a somewhat irregular basis, and sometimes the underlying concept must be deduced from actions directed. This is the single most important improvement I would recommend. The discipline of writing the cause-and-effect concepts will focus the efforts of leaders. The clarity of such concepts will explain to government employees and others why they are taking the actions directed. The connectivity of concepts between strategies would provide a narrative for the American public and the international audience. No single action in the war on terrorism is more important than improving this focus.
3) Are appropriate programs and resources provided for implementation? The identification of specific programs for execution is a strong point of this family of strategies. In fact, the documents lend themselves to use as checklists in evaluating action and progress. But the issue of resources is not adequately addressed in these strategies. Obviously, doing so would be difficult and risky. Political opponents will be tempted to take any figure as a target, arguing that it is either too high or two low. However, NSC-68 did not lay out specific spending targets – it just determined that the US would spend whatever was necessary to contain and thereby destroy an ideology hostile to its survival. The Bush family of strategies suggests a war to the death with "Terrorists with Global Reach." But it has capped homeland security spending at about the current level for the federal government and is resisting additional spending on the military

The goal is clear and admirable: win the war, secure America, manage the world to reduce dangers, all at minimum cost, so capital can remain available for investment to spur prosperity. This will require us to prioritize. And the American people must understand that our goal is to secure the nation, not every citizen in that nation. Perfect security is impossible: some civilian casualties will occur in this war. This statement is missing from the current family of strategies.
4) Is the enemy considered? Is there a way established to periodically review whether we are being strengthened and the enemy weakened? Is the strategy periodically adjusted as a result? It is probably too early to complete this analysis at this time. The new organizations required by the new strategies are still being formed. New budgets are not complete. New programs are still in development. We will probably have to wait until the end of the budget cycle after the current election year to really evaluate the impact of the strategies on the bureaucracy at all levels. There has been an effort to encourage flexibility, "red teaming," and periodic review in several areas. In others (cyber strategy comes immediately to mind), the enemy is almost wholly disregarded, on the theory that "whatever malicious can be done, some malicious person will do." This does not help in setting priorities. What the strategies probably need at this point is a strong reminder of the importance of this point – and perhaps a bit of assistance from Congress in making this an area of review and oversight. Ad always, simply highlighting the point will help those trying to turn theory into reality.
5) And finally, is the strategy designed to work over time? The answer to this question is a resounding "Yes." In fact, every strategy takes into account the danger of changing the basic nature of America in order to save it – and makes it a point to warn practitioners on this point. Like NSC-68, this family of strategies, and the concepts they represent, are intended to outlive any specific administration, and guide US efforts for the foreseeable future – or until the new threat to our survival is diminished or destroyed.

It is almost impossible to reduce the evaluation of so many strategies responding to such a complex situation to a simple "thumbs up or thumbs down." The whole point of this evaluation has been to provide a framework to recognize the subtle nuances that can mean the difference between victory and defeat in a clash with a thinking enemy.

But since an overall analysis and overall evaluation calls for an overall conclusion, I give my overall endorsement to the family of strategies described herein, and the process that produced them -- subject to the revisions and additions noted above.

Strategic Outcomes: Possible Futures

If we employ this family of plans, properly resourcing them, and evaluating and adjusting from time to time, what will the future look like? Frankly, we can't know. I can see one of four possible outcomes:

1) The family of strategies works completely. The US leads the Global War On Terror; gains global support; Federal agencies learn to lead by information sharing, as well as incentives, standards and selective evaluations; state and local agencies carry their load, training their people, minimizing their appetite for federal funds and making good use of the limited money they receive; the private sector leans forward to cooperate, bearing its share of costs and responsibilities. Attacks are limited and unsuccessful. We make significant progress and improve the world, improving our protection at a sustainable cost while discouraging

terrorism, which falls into a long decline. The enemy abandons his beliefs, and embraces ours. We determine ourt own destiny. This is our the ideal solution, but it is at odds with the nature of humans and bureaucracies. Only exceptional leadership, oversight and transparency can get us to this end state.

2) The family of strategies works only incompletely. Some agencies and jurisdictions cooperate, but others do not, taking advantage of the system, or simply ignoring the problem and expecting others to take the initiative. Accomplishing the strategy becomes less important than muddling through. Events (and hence the enemy) drive the train. We surrender our destiny, not to hard work, but to chance.
3) The strategies prove unable to constrain the rush for money. The federal bureaucracy is slow to accept its role as leader, counselor and mediator, and exercises power instead. Congress promotes the rush for homeland security money in each district, hence undermining the national strategy and priorities. The family of strategies collapses. At every level – federal, state, local, private, and individual -- a national version of "every man for himself" takes over.
4) Lack of Congressional, state, local, and private cooperation dooms federalism . Dangers demand action. Federal bureaucracy takes control of many aspects of our lives. A Homeland Security Industrial Complex arises, much as Eisenhower feared. And we become an easy mark for outside enemies seeking to weaken our government, our economy, and our nation.

The most certain thing we can do to help the administration achieve outcome #1, and avoid the others, is to publicize the strategies, hold those pursuing the strategies accountable, and support the administration in accomplishing the strategies . . . while avoiding constant intervention, and meddling on minor points. Congress has a key role here. Let the administration lead, but provide continuing oversight – as this committee has done. Adopt a congressional strategy to help the family of strategies work.

Policy makers are sometimes contemptuous of strategy, and lawmakers are sometimes too anxious to intervene in policy at the expense of strategy. Both groups should take a deep breath. Strategy determines not only how well we address the enemy in the short term, but how well we remain who we are in the long term. This family of strategies provides a good start. Give it a chance. And continue to watch it closely.

End

Mr. SHAYS. I just want to say that one of my disappointments is that we haven't truly had the kind of debate that can bring both parties together to establish what should be that bipartisan strategy. Very interesting. Thank you for the indulgence of the committee to make that comment.

Colonel Larsen.

Colonel LARSEN. Mr. Chairman and members, thank you for the opportunity to provide my assessment of these strategies. I looked at six. I didn't look at money laundering or the classified military strategy. As I said in my prepared statement, I taught strategy at the National War College, and we always told students how important the strategy is, but also, how difficult it is to develop in this town. Plans, which we heard a lot of this morning, and spending programs are easier to write and understandably so, strategy is difficult. Therefore, sometimes we end up with what the chairman refers to as ready, shoot, aim.

Looking at the six strategies, I thought there were some good plans in there. What I thought was missing was a single unifying theme that integrates all missions that were talked about this morning from deterrence, prevention, preemption, to incident management, and all participants. That is what is so different; from the President to the police officer, from a Member of Congress to a mayor, from a Cabinet Secretary to a soldier, a public health officer and a corporate CEO. That is what we do not have. Some would say that's not possible today. I disagree, and I think the members of the panel would disagree with this also.

In 1947, it has been mentioned, George Kennan gave us a single word and a philosophy behind it called containment. That guided eight Presidents, Republican and Democrat, and 20 Congresses through 40 years. I think that is what we need. We must look a little bit before we talk about the strategy that I will propose at three things strategists all look at. We understand here how the ways and means have changed, from the FBI going from reactive to proactive; how we want to exchange more intelligence information; reorientation of the military's capabilities. When we saw a soldier, an Army sergeant, ride into battle on horseback with a GPS receiver and a satellite radio, and he's guiding a B–52 designed for nuclear warfare to drop a 500-pound bomb on a machine-gun nest, we understand the ways and means have changed.

How about the end state? That's the difficult thing. We understood the end state when the Japanese bombed Pearl Harbor. Unconditional surrender. That was it. We understood the end state when dealing with Nazi Germany. We understood the end state for the cold war. What we have to do is really truly admit to the American people there is no end state. As Dr. Kass and Dr. McIntyre said, this isn't going away. If we kill all of al Qaeda tomorrow; technology will allow the other small actors to threaten us.

I used the example the president of the American Medical Association in 1967. From the scientific community the president of the medical association said in 1967, "we will soon cure infectious disease" because of vaccine and antibiotics. Almost seems humorous now, doesn't it? But there is a good lesson there, because we are curing some diseases. Within 2 years, polio will be eradicated from the human species, but we know all infectious disease won't. We may eradicate al Qaeda; terrorism we cannot. We have to learn to deal with it.

Therefore the strategy that I think provides the single unifying strategy to those six that I looked at, for that single unifying strategy, I recommend five points: one, relentless pursuit on a multilateral basis when possible of individuals and organizations who threaten our homeland; two, aggressive programs that prevent the proliferation of weapons of mass destruction, particularly nuclear and biological weapons—investments in programs like Nunn-Lugar are some of the best investments we can make; three, concentrated efforts to win the war of ideas that we have been talking about here, and some of those war of ideas are inside the United States preparing the American people—the five-step program in Israel for counterterrorism, step 5 is prepare the public psychologically. I'm not sure we're doing that; fourth, development of standards. And I know the chairman and this committee has been working on this since before September 11. We must have standards for prevention, mitigation and incident management that are fiscally sustainable for the long haul.

And now I said how good Nunn-Lugar was. Let me tell you how poor Nunn-Lugar-Domenici was. Remember the 100 largest cities? We went out there and threw all kinds of money at them, and it made us feel read good. There was no continuation training program. Colonel McIntyre and I spent 60 years in the military. You train a sergeant to fire an M–16 today, you better be prepared to train him next year or he's not going to hit anything. So we went all that money on first responders, but their turnover rate is 22 percent a year. You have to provide programs that are sustainable.

And finally, understanding that overreactions by Congress and the administration could cause more long-term damage to the American economy than the terrorists, we must be able to contain ourselves and our responses.

So the strategy that I offer that unifies these six strategies that the administration has produced, my offer is, to borrow a word from the cold war, containment. We must contain the capabilities and global reach of the terrorists. We must contain the proliferation of weapons of mass destruction, particularly nuclear and biological. We must contain the spread of hatred with an offensive campaign of our own in the war of ideas. We must contain our vulnerabilities. And we must seek to contain our response to overreact, our tendency to overreact.

This is a realistic strategy. It's one that will work, and it's one we can afford. It's a strategy that provides guidance for action and spending, and it's a strategy that's attainable and affordable, and containment is a strategy and the end state we seek.

Mr. SHAYS. Thank you very much. Thank you.

[The prepared statement of Colonel Larsen follows:]

Statement of
Colonel Randall J. Larsen, USAF (Ret)
Founder and CEO Homeland Security Associates, LLC
February 3, 2004
House of Representatives
Committee on Government Reform
Subcommittee on National Security, Emerging Threats, and International Relations

Mr. Chairman and distinguished members, thank you for the opportunity to provide my assessment and comments on America's strategy to defend our homeland.

While serving as the Chairman of the Department of Military Strategy and Operations at the National War College, I taught America's future national security leaders that a well-defined and clearly articulated strategy was the key to success. However, as a realist who has spent many years inside the beltway, I also told my students that it is quite common in this town for leaders to confuse plans and spending programs with strategies. Perhaps this is because plans and programs are far easier to write than national strategies. And frankly, there are many in this town who say programs are more important than strategies. I disagree.

It is, unfortunately, all too common in American politics to spend first and ask questions later--the DC version of "ready, shoot, aim." Consider the facts. America has been spending considerable sums of money on homeland security since 1996, but the National Strategy for Homeland Security was not published until the summer of 2002. And some, including the principal author of that document publicly admitted that is was closer to a plan than a strategy. In the words of Secretary Ridge, "It gives us a list of things to do." It was a useful document, but it did not provide the strategy that so many of us had been awaiting.

In preparation for this hearing, I examined six strategies published by the Bush Administration since the summer of 2002: The National Security Strategy of the United States of America, The National Strategy for Homeland Security, The National Strategy for Combating Terrorism, The National Strategy to Combat Weapons of Mass Destruction, The National Strategy to Secure Cyberspace, and The National Strategy for the Physical Protection of Critical Infrastructures and Key Assets. These are all useful documents. Some provide strategies for certain sectors, and most provide good plans. However, none provide a national strategy for defending the American homeland that is all-encompassing in terms of missions and participants.

That is what is missing, a single unifying theme that integrates *all missions*--from deterrence, prevention and pre-emption, to incident management and recovery, and *all participants*--from the President to the police officer, from Members of Congress to mayors, and from a cabinet secretary to a soldier to a county public health officer, and a corporate CEO. That is what's missing, the single thread that ties this all together.

Some would question whether such a strategy is possible, or useful. I will tell you it is possible, it would be useful, and there is certainly precedent. In 1947, George Kennan provided America with a strategy that guided eight Presidents, twenty Congresses, and ultimately provided victory in the Cold War. It was a strategy that could be boiled down to a single word: containment. That single concept, and the philosophy behind it, guided policy and spending programs for forty years. Today, no one has yet to offer a single unifying strategy for the challenges we face.

Of the six documents I examined for this hearing, only two provide major elements of a single unifying strategy for securing the American homeland: The National Security Strategy of the United States of America and The National Strategy for Combating Terrorism.

The key elements from the National Security Strategy are:
- Disrupt and destroy terrorist organizations
- Wage a war of ideas to win the battle against international terrorism
- Protect against and deter attack

The key elements from The National Strategy for Combating Terrorism are:
- Defeat terrorist organizations of global reach
- Deny further sponsorship, support, and sanctuary to terrorists
- Diminish the underlying conditions that terrorists seek to exploit
- Defend the United States, our citizens, and our interests

I endorse the key themes of both strategies. They both place a higher priority on taking the war abroad, rather than focusing on defense within our borders. They both address the issue of fighting a war of ideas and conditions that can be exploited by terrorist organizations and both talk of protecting and defending the homeland. These are most certainly elements of a homeland security strategy, yet they do not provide a single unifying strategy.

Many question whether it is possible to develop a concise yet broad strategy such as *containment* in the Cold War or *Europe first* in World War II. However, I believe it is possible and I believe the American people, and particularly the 10 million Americans directly associated with homeland security deserve such a strategy.

Let's face it, how many in this town and this nation have read the six documents mentioned above? Not many I suspect. But if few have read all of these documents, how can they successfully develop and implement plans and programs to defend our homeland? It would be like going to the Super Bowl without a game plan.

To design a single strategy for homeland security, one must begin with assumptions, and these assumptions are far different from the Cold War, or perhaps, any other time in our history. Strategists talk of *ends, ways and means.* Most agree that the *ways* and *means* have changed dramatically. During the Cold War, preemption was considered taboo, because it was a euphemism for first use of nuclear weapons. Whether or not you agreed with the President's decision to oust Saddam Hussein from power in Iraq, preemption is clearly an option of American security policy in the 21st century. In Afghanistan, it was an Army--that for decades had prepared for large tank battles in central Europe and the deserts of Southwest Asia--found its soldiers riding into battle on horseback, using laser designators and satellite radios to guide 500 pound bombs being dropped from airplanes built in the 1960s to fight a nuclear war. The *ways* and *means* have definitely changed. I am not, however, sure, that most understand the change in the *end-state*.

When America entered World War II, we understood that Nazi Germany and Imperial Japan could be defeated. When the Cold War began, we believed that a containment of Soviet expansion would eventually lead to the collapse of the Soviet Empire. But who today truly believes we can defeat terrorism?

In 1967 the President of the American Medical Association stated that the end of infectious disease was possible through the use of vaccination and antibiotics. Obviously he was mistaken. While it may be possible to eradicate some infectious diseases, just as it may be possible to eliminate al Qaeda, winning the war against terrorism is as likely as winning the war against infectious disease. The best we can realistically hope for is to contain the frequency and severity.

A strategy to defend the homeland is far more complex than winning the war against al Qaeda. We must understand this is about a permanent change in the international security environment. We must think long-term and we must seek an end-state that is realistic. The technological genie is out of the bottle--small actors can now threaten a super power. **This fact will not change.**

Therefore, a single unifying strategy for defending the American homeland must contain the following elements.

- Relentless pursuit, on a multilateral basis when possible, of individuals and organizations who threaten our homeland … this includes those who support them
- Renewed and aggressive programs to prevent the proliferation of weapons of mass destruction, particularly nuclear and biological weapons
- Concerted effort to win the war of ideas, particularly important in the information age
- Development of standards for prevention, mitigation and incident management programs that are fiscally sustainable for the long-haul
- Understanding that over-reactions by Congress and the Administration could cause more long-term damage to the American economy than terrorists

For more than two years I have been searching for a single word or phrase that could capture these four elements. The single word capable of providing an overall strategy for defending the American homeland is not new. I borrowed it from 1947 and George Kennan, however, the philosophy behind the strategy of *containment* in the 21st century is far different.

It is unrealistic and even naïve to believe that we can permanently end terrorism or terrorist threats to our homeland. One of the candidates for President recently stated in a television advertisement that he could prevent attacks on the American homeland--a preposterous idea that he quickly withdrew. Nevertheless, in the case of defending our homeland, we all hate to admit that which is true. We cannot defeat terrorism. We cannot win the War on Terrorism.

Unconditional surrender by the Germans and Japanese ended the threat. That is not possible today. Secretary Ridge has stated that there will be no victory parades. He is absolutely correct. Therefore, let us make our strategy reflect this reality. We should seek to control certain factors, or better yet, *contain* the threat from terrorism.

We must *contain* the capabilities, global reach, and financial resources of terrorists and terrorist organizations. We must *contain* the proliferation of weapons of mass destruction, particularly those weapons that most threaten our survival, nuclear and biological. We must *contain* the spread of hatred with our own offensive campaign in the war of ideas. We must *contain* the vulnerabilities of this nation. And we must seek to *contain* our response to these new threats. We must not overreact.

Some will comment that this is a defeatist strategy. I say it is realistic. We cannot stop every determined truck bomber, but we must prevent a mushroom cloud over an American city or a catastrophic biological attack on the nation. We can't kill, capture, or deter every terrorist, but must *contain* them by limiting their capabilities, their global reach and financial resources.

We cannot prevent the proliferation of all weapons of mass destruction. Chemical agents, including industrial chemicals are far too easy to produce or buy. Radiological material for use in a dirty bomb has already proliferated beyond control. It exists in most hospitals, laboratories, and even at many large construction sites around the world. However, we must *contain* the proliferation of nuclear weapons and biological weapons. Programs such as Nunn-Lugar are great investments in homeland security.

The Wahabi sect of Islam supports schools, organizations, and special programs (some in our own country, particularly in our prisons) that are registered with the IRS as 501 (c) 3 charitable institutions that preach hatred and violence against America and Americans. We cannot end all coordinated information campaigns against the US, but we must retaliate with our own offensive campaign to *contain* this contagion of hatred, disinformation, and instigation.

We are a free and open nation. That makes us a target rich environment for terrorists. We must take prudent and fiscally responsible action to reduce these vulnerabilities and implement realistic and measurable prevention and incident management programs. The measurement part is critically important. If we don't set standards and goals, how can we measure progress?

One distinguished group of Americans released an often quoted report last year calling for an *increase* in spending on security within US borders that would approach $100 billion over five years. But we have yet to establish standards and measurable goals for such programs. How did they determine these numbers? How would Congress allocate and prioritize spending? It would be a great for pork. It would send money to every Congressional district. But would it make us more secure?

The press has a field day when a college student smuggled a few box cutters on an airliner, but do we really want a security system that is 100 percent successful? If so, it will take us hours to get through an airport. A system that is 80 percent effective is not an attractive target--even to a suicide bomber. A system that stops four out of five attackers is a strong deterrent, a system we can afford, and if it is part of a layered defense, it will provide the security required. A passenger and cargo screening system, backed up by hardened cockpit doors, thousands of armed sky marshals, armed pilots, and passengers who have not forgotten Todd Beamer and his compatriots is the type of security system we need and can afford.

Finally, we must not allow Congress or the Administration to overreact. This will be most difficult during election years. On some days, the hyperbole, hype and hollow promises of some politicians frighten me more than terrorists. Following the President's State of the Union address, a prominent Democratic leader stated that less than five percent of cargo entering the US is currently inspected. She demanded that 100 percent of cargo that comes into this country by sea, and 100 percent of the cargo carried on domestic and international flights be inspected. That is a recipe for economic disaster. That is what I mean when I say the US government could do more damage to the American economy than terrorists.

It is important that I maintain my nonpartisan status, so let me go on the record that I have heard equally troubling statements from Republicans, such as spending billions of dollars securing our borders. According to the Department of Homeland Security, there are 7,000 miles of borders and 95,000 miles of shoreline in this country. Understanding that we are in this for the long-haul, how could we ever hope to seal these borders against terrorists? Imagine the costs. It is not economically feasible. We must *contain* our impulse for overreaction. Programs such as these will make us no more secure and divert money away from programs that could. This tendency for impulse spending and regulation will be most likely during election years and immediately following attacks.

And yes, there will be more attacks. We must never forget the words of Ramsey Yousef, the mastermind of the 1993 bombing of the world Trade Center. After his arrest in 1995, he was being flown into New York City for arraignment. John O'Neil, the FBI's Chief of Counter Terrorism pointed to the World Trade Center towers and said, "They are still standing." Yousef answered with, "We are not done yet."

Mr. Chairman, al Qaeda is not done yet, and more importantly, we need to understand there are others out there who will one day follow in al Qaeda footsteps. We are in this for the long-haul. We must have a single unifying strategy that responds to the realities of the 21st century.

Containment is the strategy that provides the common thread to all others associated with defending the American homeland. It is a strategy that provides guidance for actions and spending. It is a strategy that is attainable and affordable. *Containment* is both the strategy and the end-state we seek.

Mr. SHAYS. Mr. Cilluffo.

Mr. CILLUFFO. Thank you, Chairman Shays and distinguished members of the committee. It's good to be back and in familiar surroundings to discuss our strategies to combat terrorism and secure the homeland. Like Dr. Kass, my insights or thoughts are my own and obviously do not reflect my views and my time at the White House and/or other organizations, the Homeland Security Advisory Council and others I may be part of. Given time constraints, I will try to be brief. Not one of my strong suits.

Mr. SHAYS. Be concise.

Mr. CILLUFFO. I will deviate from my prepared remarks and highlight a few of its key points.

Like Dr. McIntyre, I would like to compliment the subcommittee for its leadership and longstanding role in helping frame and shape the strategies before us today, and also for recognizing that we cannot march into the future backward fighting yesterday's wars alone. We need to remember that September 11—the attacks of September 11 were not merely a snapshot in our Nation's history. We are in a new normalcy now. The threat remains very real, but yet may come at us in various forms and ways and in morphing ways. This living agile enemy bases its actions on our actions, seeking out and exploiting our vulnerabilities. Thus we must be willing to learn from our successes and mistakes and effectively manage risk by constantly reevaluating our policies and recalibrating our programs in order to stay ahead of the terrorists.

In order to combat these ambiguous and moving targets, we need a national strategy that is flexible, comprehensive and coordinated; living strategies, if you will. From my perspective, the President acted decisively on this need. In conjunction with one another, the strategies before us today provide a comprehensive national strategy to win the war on terrorism on all fronts.

A comprehensive strategy to combat terrorism must employ every instrument of statecraft to attack the enemy on all fronts and secure our homeland. For example, you cannot separate homeland security policy from economic policy from foreign policy from national security policy from military policy from health policy from science policy and technology policy. It is messy, and I think Congress realizes it's messy, in terms of trying to get your arms around this challenge. It is cross-cutting by its very nature, and they are inextricably interwoven, and you cannot treat policies in isolation. It's not about building a little black box that says break glass when something bad happens.

I love the term that Mr. Yim used earlier. It is about embedding tactics, operations and existing tactics and operations, and it is about integrating a whole wherein the strategies feed off and enable one another.

The task of securing the homeland has been cast by some as a choice between security or freedom or security or competitiveness. We heard the discussion earlier today. These are not mutually exclusive propositions. In fact, we can and we must have both. The single tenet that underpins everything we are doing, it is not about security or freedom, it is about securing freedom. And we can never forget that. And we need to do so in a way that projects our values.

We need to protect Americans, but we always need to protect and project America.

The overall strategy to combat the threat of terrorism must incorporate the marshalling of these domestic resources with the engagement of the international allies and assets. We should learn from the experience of our allies. Many have had decades of terrorism that they have had to deal with over the years, and we should continue to build on some of the successes that we are learning as we are prosecuting this war and as we are moving into it day in and day out.

I think the National Strategy for Combating Terrorism recognizes that we also need to be proactive and extend our defenses outward. We discussed earlier some of the questions raised by some of the Members here. What are some of those specific international issues we need to be able to address? And quite honestly, we want to be able to push the border out, widen the net to stop terrorists over there, and not waiting until they reach our shores right here. And to do this, we need to recognize that a transnational threat will require transnational solutions. We need to maintain a coalition of countries dedicated to isolating not only terrorist organizations, but also the nations that sponsor, support or harbor them. And I think the National Security Strategy of the United States makes that clear.

Bringing all the instruments of statecraft to bear will not only pressure these countries to cease actively or passively harboring terrorist organizations, but also pressure them to take the initiative to deal with the terrorist problem within their own borders and ultimately drain the swamp that spawns terrorism. I clearly see that as one of the end states.

Let me just say a brief word because both Congressman Schrock and Congressman Platt brought this up earlier about intelligence. It is the life blood of the war on terrorism whether in support of diplomacy, covert action or in support of military, law enforcement or homeland operations. Intelligence not only provides the detailed information we need to preempt attacks, seize terrorist assets and identify terrorist capabilities, it can also provide us insight into what the terrorists value, allowing us to go on the offensive and take it away.

It is critical to illuminate key vulnerabilities that can be exploited and leveraged to preempt, prevent and disrupt terrorist activities before they occur. And I think that the mix between signal intelligence and human intelligence was one that we did for years, Congressman Murphy, neglect. I think that is slowly changing, but you have to realize it takes time. You don't push a button, and it is not as easy as knocking on bin Laden's cave and saying, hi, I am here to join. This is going to take years potentially to get that right. But clearly the objective should be to get there before the bomb goes off.

We want to be able to fragment the adversary, to fragment its enterprise and attack the pieces, which I think is one of the action plans we have been working toward. That said, we can never guarantee with 100 percent success in preventing all attacks. Immediately following September 11, the President led an assessment to identify what policies, programs, procedures worked, which didn't, and what are the major gaps and shortfalls that needed to be backfilled. In a way, we were building an airplane midflight.

As we go about culminating in the President's National Strategy to Secure the Homeland, we are also going through the greatest transformation in the Federal Government's history since the National Security Act of 1947. Dr. Larsen mentioned the containment word. If I were forced to put our homeland security strategy on to a bumper sticker, that word would be to connect; to first connect the many Federal departments and agencies that have a role in securing the homeland. The President came to the conclusion that the whole was less than the sum of its parts; hence the creation and marrying up of authority, accountability and resources with the new Department of Homeland Security. But it also meant identifying who needed a seat at the national security planning table. This isn't just the regular suspects, FBI, CIA, Department of Defense. Primary care physicians, entomologists, agricultural services inspectors, people who have never really been part of the national security community not only needed a seat, but a front-row seat.

Culturally there are huge challenges. One community wanted to string them up—law enforcement, the other community, string them along—intelligence, and then you got the health component that just wanted to deal with the strung out. Very different views on the world. So we want to be able to bring some of these capacities together.

But the Federal piece is easy compared to interfacing with Federal, State and local. Obviously any national strategy needs to be national, not Federal. And we all know that those first to arrive and last to leave will be our Nation's emergency responders. They are the ones who need the tools, the capacities and the wherewithal and will ultimately determine whether or not the battle can be won or lost.

We discussed the private sector. They own and operate a majority of the infrastructure. This can't be a "thou shalt" from Washington. It needs to be a partnership—work with. I personally believe it should be mitigate before litigate or regulate, but we need to be able to put some pressure on some of the shared responsibilities of the private sector, and it is a shared responsibility. Government needs to lead by example, get its own house in order, and only then can they expect the private sector to do the same.

Congressman Tierney, you mentioned the American people. I think this is a primary tenet of the national strategy. We need to get information to citizens on what they can do to protect their families and their communities; the Citizen Corps, part of USA Freedom Corps, the "ready" campaign asked people to start thinking not to ask how afraid should I be, but what can I actually do about it. And the President's view was the best way to defeat evil is to do some good and to reinvigorate some of the public service that is available.

Let me also——

Mr. SHAYS. No. Let's close up here.

Mr. CILLUFFO. Let me close very briefly to state—and I will use the wise words of Yogi Berra, who I consider one of the greatest strategists and philosophers: The future ain't what it used to be. And the best way—and I think it is also fair to say that since the end of the cold war, threat forecasting has made astrology look respectable.

[The prepared statement of Mr. Cilluffo follows:]

THE GEORGE WASHINGTON UNIVERSITY
WASHINGTON DC

OFFICE OF THE ASSOCIATE VICE PRESIDENT
FOR HOMELAND SECURITY

Combating Terrorism:
Developing Effective Strategies Against Terrorism

Statement of Frank J. Cilluffo

Associate Vice President for Homeland Security
The George Washington University

Before the U.S. House Committee on Government Reform
Subcommittee on National Security, Emerging Threats,
and International Relations

February 3, 2004

Chairman Shays, distinguished members of the committee, it is a privilege to appear before you again today. In holding these hearings the Committee on Government Reform, and Congress as a whole, should be commended for its continuous efforts to evaluate how our current policies and programs come together and to identify gaps and shortfalls within them so that they may be remedied to enhance the security of our homeland. This subcommittee in particular, should be proud of its longstanding role in framing and helping shape the national strategies to combat the threat of terrorism before us today. It is only with efforts like these that we will be able to continually develop, integrate, and implement effective "living" strategies, which are vital to combating this dynamic threat.

The September 11th attacks were not a "snapshot in history." We are in a new normalcy where the responsibility for protecting the homeland from terrorist attack remains will be with us now and well into the future. We must remember that we do not face a single, geographically anchored enemy but a myriad of threats, smaller in magnitude and harder to see and counter. A successful overall national strategy to combat these ambiguous, amorphous, moving targets must be flexible, comprehensive, and coordinated. It is with this recognition that the current strategies for securing the homeland were created.

The President has acted decisively on the need to have an integrated overall strategy to combat terrorism. In the weeks following September 11th, the President issued a directive that tasked the government to direct every resource at its command—all tools of diplomacy, intelligence, law enforcement, and financial influence—to win the war against terrorism; the President has led the way to ensure the directive was acted upon. Less than three years ago we did not have a comprehensive strategy for combating the threat of terrorism or the substantial challenges of homeland security. Now, under the President's leadership, we not only have the *National Strategy for Combating Terrorism*, the *National Strategy for Homeland Security*, and the *National Security Strategy of the United States*, but we also have the *National Military Strategic*

2300 I STREET, NW • SUITE 721 • WASHINGTON, DC 20037
202-994-0295 • FAX 202-994-2543 • www.homelandsecurity.gwu.edu

Plan for the War on Terrorism which provides a clear framework for how the U.S. Armed Forces continue to conduct the war on terrorism and the *2002 National Money Laundering Strategy* which is the first to outline a government-wide strategy to combat terrorist financing in order to destroy the conventional and unconventional financial tools on which the enemy depends. In addition, the President has provided us with essential guidance to address specific concerns including the *National Strategy to Combat Weapons of Mass Destruction,* the *National Strategy to Secure Cyberspace,* and the *National Strategy for the Physical Protection of Critical Infrastructures.* All of these documents, in conjunction with one another, provide the comprehensive national strategy we need to win the war on terrorism on all fronts. Terrorists are seeking to exploit our vulnerabilities—these strategies provide a clear way forward to prevent them from doing so while protecting that which we hold dear.

On September 11th, the terrorists attacked highly visible symbols of our military strength and our economic prowess. Though exceedingly well planned, coordinated, and executed, the comparatively low-tech means employed by the terrorists raises the future possibility of a well placed bomb or attack meant to cause mass effect; a chemical, biological, radiological, or nuclear (CBRN) attack; a cyber strike; or a more inclusive, more sophisticated assault combining both physical and virtual means on one, or several, critical infrastructures. The threat remains very real. A low-tech, high-tech combination attack is an especially dangerous possibility, for while Bin Laden may have his finger on the trigger of an AK-47, his nephew may have his finger on a computer mouse. Such a scenario demonstrates the need for an integrated, comprehensive approach rather than one that tries simply to isolate and counter a single threat.

Thus, a comprehensive strategy to combat terrorism should incorporate a full spectrum of organized actions by employing every instrument of statecraft to attack the enemy on all fronts and secure the homeland. Homeland security policy is inseparable from economic policy, health policy, national security policy, and foreign policy—all of which must exist underpinned by the rule of law. The task of securing the homeland has been cast by some as a choice between security *or* privacy, security *or* freedom, and security *or* competitiveness. These are not either-or issues; we can and must have both. We cannot codify our activities into neat, clean boxes or treat elements of the strategy isolation; this threat requires a balanced and integrated approach.

Accordingly, the overall strategy to combat the threat of terrorism must incorporate the marshalling of these domestic resources with the engagement of international allies and assets to be effective. To truly defeat terrorism, we must be cognizant of the fact that this is a transnational threat that requires transnational resources and solutions. The shift away from political and towards ideologically based terrorism means that many more countries have become direct targets of escalating acts. As a result, many countries now have a vested interest in studying and defeating terrorism. Indeed, some already possess a breadth of knowledge and experience from dealing with years of terrorism within their own borders. We should learn from the experiences of our allies, and build on the successes we have had thus far in prosecuting this global war on terrorism.

The *National Strategy for Combating Terrorism* recognizes that the war on terrorism cannot be won without employing resources abroad in collaboration with our allies. It also makes the important point that in order to *defeat* terrorist organizations of global reach, *deny* further sponsorship and support, *diminish* the underlying conditions that terrorists seek to exploit, and *defend* the United States at home and abroad, we need to be proactive in our efforts by extending our defenses outward. This means stopping the terrorists abroad before they ever reach our shores. Relying on catching the terrorists at our borders is not enough to protect the homeland. We must push the protection of our borders out—widening the net to catch the terrorists. To do this we need to continue to maintain a coalition of countries dedicated to isolating not only terrorist organizations, but also the nations that support or harbor them.

In the *National Security Strategy of the United States*, the President recognized this need, vowing to hold accountable nations that are compromised by terror including those that harbor terrorists or support terrorism. These countries still pose a significant threat to the United States because they can share information, technologies, means, and capabilities with terrorists. We need to continue to work cooperatively when possible to use all of the tools at our disposal including law enforcement instruments to prevent such transfers, military instruments including covert action to preempt imminent attacks, economic instruments to starve the terrorists of funding and punish those who provide financial support, and diplomatic instruments to isolate nations that harbor terrorists. The consequences of harboring terrorists should be made too great for a nation to consider it acceptable. Bringing all these instruments of statecraft to bear will not only pressure these countries to cease actively or passively harboring terrorist organizations, but also pressure them to take the initiative to deal with the terrorist problem within their own borders. We can offer support to those countries that continue to join our coalition and commit themselves to fighting terrorism by helping to train and equip their indigenous authorities so that they can drain the swamp of terrorism. But in order to know what clandestine activity these nations are involved in and apply pressure for them to cease their support of terrorism and join the coalition, we must refine the most important tool we have to combat terrorism—Intelligence.

Underpinning every aspect of the war on terrorism is the need to have a first-rate intelligence capability. Accurate and timely information, coupled with proper analysis, is the lifeblood of the war on terrorism. Combating the breadth, depth and uncertainty of the terrorist threat demands significant investment, coordination and accuracy in the intelligence process across the board. The intelligence community has made great strides in information sharing and coordination among intelligence agencies and security services, but it must continue to be vigilant in its analysis to provide accurate, timely, and actionable intelligence. Every aspect of the campaign—from diplomatic efforts to covert action to financial and political operations to the provision of warnings about future attacks—relies largely on our intelligence, coupled with intelligence from allies. Intelligence not only provides the detailed information we need to preempt attacks, seize terrorist assets, and identify terrorist capabilities, it also can provide us insight into what the terrorists value, allowing us to go on the offensive and take it away. Intelligence involves understanding the motivations, thoughts, and plans of one's enemies. It is also critical to illuminating key vulnerabilities that can be exploited and leveraged to prevent, preempt, and disrupt terrorist activities before they occur. The goal here is to obtain the intelligence needed to isolate the military and operational planners from their organization, and terrorist organizations from their network in order to fragment the enterprise and attack its pieces. Ironically enough, even the vilest terrorist depends on the "honor" of another terrorist to do his or her work. Once that honor and loyalty is breached, the system of trust—the glue of the organization—collapses. In addition to illuminating vulnerabilities within the terrorist network, intelligence provides insights into the cultures and mindsets of terrorist organizations that are crucial to providing indications and warnings of possible attacks. The first priority should always be to get there before the bomb goes off; having a top-notch intelligence ability is the way we do that.

Nevertheless, no matter how hard we work and how many resources we invest to prevent another attack from occurring, we cannot guarantee 100 % success. Understanding this, the President implemented measures to protect the vulnerabilities we have at home and build up our capacity to mitigate the effects of a terrorist attack and minimize the loss of life. The President's *National Strategy for Homeland Security* provides clear goals and objectives for how this should be accomplished, linking the diplomatic and intelligence pieces together with the response needed at home. In addition, the President recognized that coordination and integration of these efforts was essential for success. To accomplish this synergy, he proposed large, sweeping actions to protect and defend the homeland—namely the creation of the Department of Homeland Security. Together, the President and Congress worked to stand up the Department, achieving the most significant reorganization within the U.S. government in over 50 years. At its creation, DHS was tasked with preventing terrorist attacks within the United States, reducing America's vulnerability to terrorism, minimizing the damage, and enhancing the response and recovery efforts should an attack occur. Under the outstanding leadership of Secretary Ridge, DHS has been working tirelessly with other agencies to analyze threats and intelligence, guard our borders and airports, protect our critical infrastructure, and coordinate emergency response efforts. The Secretary and the department deserve to be commended for what they have accomplished in such a short period of time.

But we are still in the early stages of this war on terrorism and DHS recognizes that there is much more to be done to secure the homeland. Paramount among these future actions is the need for enhanced coordination among all levels of government. We must ensure that we continue to connect relevant federal entities with each other but we also need to connect federal authorities with state and local officials, states with other states, all levels of government with the private sector, and each of these actors with the American people. Terrorism is at its very core a psychological weapon, intended to erode trust and undermine confidence in our government, its elected officials, institutions or policies. Without working relationships of trust and mutual confidence between and among all of the actors who are key to our efforts to fight terrorism, the overall strategy to prevent and prepare for terrorism will be defeated. This is why it is absolutely essential that we connect all of the relevant players in homeland security—we cannot be exchanging business cards on game day.

DHS is the belly button that links this whole system as it provides a central clearinghouse to marry up accountable resources and actors, making sure that all of those who need a seat at the homeland security table have one. This is an especially important function for DHS in fostering a healthy and reciprocal public-private partnership. The vast majority of the owners and operators of our critical infrastructure are in the private sector. And, as the *National Strategy for the Physical Protection of Critical Infrastructures* and the *National Strategy to Secure Cyberspace* emphasize, critical infrastructure protection is a shared responsibility that cannot be accomplished by the government alone. But the government must lead by example—getting its own house in order—and then driving the guidelines and best practices for the private sector. By then building the business case for homeland security, the government will foster the public-private partnership. This will require coordinated action on the part of federal, state and local governments, the private sector, and American citizens to secure the infrastructure from virtual or physical attack.

Securing the homeland relies on the very essence of federalism. This principle is embodied in the cooperation required for critical infrastructure protection, but it is also manifested in the communication now occurring between the federal government and state and local emergency responders so as to ensure seamless coordination between state and local emergency personnel and federal assets. They are continually working to clearly allocate between and among one another the responsibilities and resources for emergency preparedness and response while making a concerted effort to ensure the harmonization and interoperability of equipment and incident command structures. Such organization and coordination figures most prominently in the area of emergency preparedness and response, particularly when responding to a catastrophic CBRN attack. The government must be able to adapt to, cope with, and manage the myriad of multi-dimensional issues that CBRN terrorism poses. The *National Strategy to Combat Weapons of Mass Destruction* set forth the urgent objective of developing and maintaining the capability to reduce the horrific consequences of such catastrophic attacks. But as it stands now, the medical and public health communities would be severely strained in the case of a CBRN attack, particularly with the challenges of bioterrorism. To address this, we must continue to enhance the core capacity for public health and medical preparedness.

The President has been working to this end. Recently, DHS, along with the Department of Health and Human Services announced that the President's FY '05 budget request will include a $274 million Bio-Surveillance Program Initiative that will provide some of the improvements needed in this area. The initiative will build upon the on-going BioWatch program by enhancing surveillance in human health, hospital preparedness, state and local preparedness, and vaccine research and procurement, with the overarching goal of integrating all of these surveillance efforts across the government into one comprehensive system. The tools of epidemiological surveillance and detection that it calls for are vital to protecting the homeland from the very real and very deadly threat of bioterrorism. This initiative embodies the integrative approach we need to have in combating terrorism throughout all levels of the government.

To complement the strides we have made and are continuing to make in the surveillance arena, we also need to make progress on the President's Project BioShield. I applaud the House of Representatives for passing HR 2122, Project BioShield Act of 2003, and I encourage the Senate to do the same. Project BioShield will give us the tools we need now to bring the best and the brightest of researchers, medical experts, and the biomedical industry together to develop more effective vaccines and countermeasures to protect against biological warfare agents. The President has introduced these vital programs to protect against one of the world's most dangerous threats, but in order to link together these important bio programs, it may be necessary to add another national strategy into the mix. Namely, an end-to-end strategy to combat bioterrorism—from prevention through treatment—by better integrating the bio-medical industry, our nations hospitals, healthcare providers, physicians, agricultural services inspectors, and entomologists, to name a few. In our efforts to secure the homeland against bioterrorism and a plethora of other threats, both our capabilities and organizations must continue to be strengthened, streamlined, and synergized so that effective prevention will enhance emergency preparedness and response and vice versa.

It is true that as many resources as the government devotes to protecting the homeland, it is not possible to protect against everything, everywhere, all the time from every adversary and every modality of attack. Our resources are finite. This is why it is critical to continue to prioritize resources, generating a national return on our investment by identifying initiatives that will maximize secondary and tertiary benefits beyond guards, guns, and gates. Strengthening the ability to deal with the extraordinary (i.e. bioterrorism) provides tools and capabilities that are equally valuable in dealing with the ordinary (i.e. the flu).

Still, the task before us remains enormous. This mission of securing the homeland is much like the role of a goalie in a hockey game. The goalie does not have many opportunities to score a goal, but when his team's net is threatened, it is imperative that he be successful in blocking the attack. To prevent an attack, we've got to be right every time, all the time, whereas the terrorist needs only be right once to succeed. This is why it is so important for us to train and exercise to continually test our preparedness and response. We want the mistakes to be made on the practice field, not on game day on Main Street in Somewhere, USA.

General Eisenhower once said that in preparing for battle plans are useless, but planning is indispensable. It is in the planning, organizing, training, and operationalizing of our national strategy to combat terrorism that we will win this war. The development of these strategies to date has provided much-needed guidance in the almost two and a half years since September 11th. But we are still in the early stages of war. The old military adage goes like this: Amateurs talk about strategy; professionals talk about logistics. We have a national strategy before us that is working. Now we need to continue to concentrate on execution. To translate the strategy from the 10,000-foot level all the way down to the ground, we must push capacity to the frontlines, to the muddy boots and white coats.

Our adversaries recognize that we cannot be defeated in a conventional war, tank for tank, plane for plane on the traditional battlefield. Thus, the terrorist enemy is employing asymmetric tactics to offset our strengths and attack our weaknesses. They're searching out our vulnerabilities. Though it is not possible to protect the homeland from every fathomable attack scenario, at least not in a democracy such as our own, we can stay one step ahead of the terrorists by keeping them on the run while simultaneously securing our critical vulnerabilities from attack. In the words of Benjamin Franklin, failing to prepare is preparing to fail. But we cannot afford to fail this test. We must think the unthinkable—because the terrorists are thinking it—and then we need to take actions to prevent it from happening while we still have time to do so.

The subcommittee is meeting today for this purpose. The overall national strategy to combat terrorism and the individual strategies under review at this hearing recognize that the crosscutting nature of the threat requires that we treat the actions the government implements as an integrated whole. These are inextricably interwoven, living strategies. But any successful strategy to combat terrorism will require continually monitoring and measuring the effectiveness ("benchmarking") of the many programs that implement it so as to lead to common and integrated standards, practices, and procedures. The terrorists want us looking over our shoulders in fear of an attack but we need to keep the terrorists looking over their shoulders, not knowing when, where, or how we will strike. We cannot march into the future backwards and fight yesterday's war alone. This living, thinking enemy bases its actions on our actions. Thus, we must be willing to learn from our successes and mistakes by constantly reevaluating our policies and programs in order to stay ahead of the terrorists, prevent future attacks, and secure the homeland.

Policy and strategy without resources is rhetoric. It is imperative that the President and Congress continue to set their sights on the comprehensive implementation of a living national strategy to combat terrorism. This process of turning concepts into capabilities will require not only vision but also sustained political will. It is the responsibility of policymakers on both ends of Pennsylvania Avenue to be enablers in marshalling and mobilizing the vast resources of the United States to combat this threat for today, tomorrow, and for years to come. We cannot be lulled into a sense of complacency. Instead, we must present a sustained, united front to defeat terrorism at home and abroad so that we may have an America that is not only more safe and secure, but better too.

Thank you for the opportunity to once again share my thoughts with you. I would be pleased to try and answer any questions you may have.

Mr. SHAYS. Mr. Schrock.

Mr. SCHROCK. Mr. Chairman, we have attended a lot of hearings, but these are probably five of the most fascinating informational people we have ever had, and we thank you very much.

Dr. Kass, you mentioned Clausewitz, which gives me goose bumps because I had to read that book on war when I was at the Naval War College, and I stuck it away, and we're moving out of our house, and I was looking at the books, and there it was. Believe it or not, I'm going to read it again, because I really believe it will apply to a lot of what we are doing here. So that is one you are right on.

You talk about patience. You talk about patience. We don't have patience in America. We want instant gratification. We thought the minute we went in and bombed Afghanistan the first day, it was over and everything was going to be fine, and that is an education process the American people clearly need to understand.

And, Colonel McIntyre, you said something I'm going to remember for a long time: managing dangers, not eliminating them. As much as Ed Schrock would like to eliminate all these dangers and get rid of these guys, I'm afraid we are not going to be able to do that. The Vice President has said if we leave one terrorist standing, they are going to put roots in the ground and continue to grow. And that is nice to think we might get rid of everybody, but if we can manage that threat, that is probably some—and know where to stop, that is a fascinating comment. I'm going to be thinking a lot about that, too.

And the strategies can change in every administration, and they do. You are starting to hear that on the campaign trail, if I am elected, I will do this, and I will take this action. And I am not sure all that is good for the long-term role or goal in trying to get rid of the terrorists.

And, Colonel Larsen, preparing the public, that is one of the hardest thing we have to do, because I think they want this thing over, and they think it's going to be over. But they need to be educated that it's going to be a long time.

And sustainable programs, you're right. It's fully funded. It's a feel-good thing. We do it. We think we are done with the job, where in 5 years everybody who was there who got the training is gone, and we need to get that up and going.

And Mr. Cilluffo talked about new normalcy. We are never going to be the same again, and that is a very, very sad thing, but we need to stay ahead of the terrorist.

I am not going to ask you the two longest questions. The coordination of the agencies is real important, and the heads of the cabinets, the butting of the heads of the Cabinet members, how do we solve this? How do we get these agencies to work together so everybody is talking off of one sheet of music, so everybody out there isn't doing their own thing? I don't understand that, and maybe you do.

Dr. KASS. The only way you can do that is exactly the way the committee is trying to do it, namely what is the overarching strategic design; what is it that we are supposed to be all trying to accomplish, and only then you can go from strategy to specific tasks that assign to the various agencies. Right now everybody is doing everything, and you have no clarity.

Dr. MCINTYRE. We are going to have to find some way to reward people. You know, when you play in the Super Bowl, you get paid more if you are on the winning team. You don't get paid extra just for being really good at defense or being a really good pass receiver. So everybody plays for the team. But our entire system is constructed for individual or agency or local evaluation and consequently local reward. We have to find a way to reward the entire system when it succeeds and punish the entire system when it fails. That is very difficult to do, but I am telling you the individual reward is not the answer to moving the team as a whole forward.

Mr. SCHROCK. What I hear you saying is that means going into these agencies and rooting out some of the mentality that's been there forever that wants the status quo and doesn't want things changed for their own security?

Dr. MCINTYRE. The single greatest obstacle we face in changing the bureaucracy is to undo the successes of the past. It is not the failures of the past, it is successes of the past is the problem, because people will continue to do that because it has been successful in the past.

Mr. SCHROCK. But what is successful in the past doesn't apply.

Dr. MCINTYRE. Our whole structure is built from our academic system forward. From the 1500's, we built an academic system that is vertical, and that is the way people are rewarded. Our problems today are horizontal.

Our problems today are horizontal, and we've got to find a reward structure that is horizontal in nature and not just vertical in nature.

Mr. SCHROCK. We will—I think the Secretary of Defense is trying to do that in his reorganization of the Pentagon.

Give us an example of how you do that.

Dr. MCINTYRE. Jointness is a very good one in that you are not necessarily promoted for being a really good Army officer anymore.

You are rewarded for being part of a joint team, for unless you have proven yourself in that joint team there is no advancement no matter how good you are in the Army or the Navy.

We are going to have more—I don't like necessarily the word "jointness" to apply but more interagency—reward for interagency behavior.

Mr. SCHROCK. Purple.

Dr. MCINTYRE. "Purple" is a good word. "Interagency," I think, is the proper word.

You do that, you know, the Congress did that with the services by making the requirement that you had to serve jointly for advancement to general officer.

When that kind of requirement becomes the commonplace within the agencies in the U.S. Government, then cooperation and interagencies will be desirable, in terms of where you send the extra person out of the office.

Colonel LARSEN. You used the term "patience" a moment ago in talking to Dr. Kass.

Remember, that took 40 years to get it right, to get Goldwater-Nichols, but it's a commitment to that long-term effort because it took Congress, not the administration or the Pentagon, to give us Goldwater-Nichols, to give us jointness so we could work together. So it is going to take action by this body and time.

Mr. SCHROCK. I agree.

Mr. CILLUFFO. Congressman, two points come to mind.

First, the Homeland Security Council, in conjunction with the National Security Council and the Executive Office of the President, does have a Deputy Assistant to the President that supports both the Assistant to the President for Homeland Security and the Assistant to the President For National Security, Dr. Rice and Dr. Gordon, but let me also say that clearly the turf we should all be worried about is the turf we are all standing on and the horizontal challenges in conjunction with the vertical challenges are not easy.

I believe General Eisenhower, and it's in the Pentagon on the way to the bubble, and it's a quote and I'll paraphrase it: In preparation for war I have found plans to be useless but planning to be indispensable, and I feel the training and exercising component of this is so important. We can't afford to exchange business cards on game day. We need to get people to be facing one another, to understand the roles, to understand their limitations, to understand what their actual missions are, at the Federal level and at the Federal, State and local level.

The words mean something very different.

Lexicon. The word "surveillance" to an epidemiologist means something very different than it does from a military perspective, from a C4ISR perspective than it does to law enforcement.

This is a transformational change that will take some time for us to get right. I'm not sure it will ever be right, but one thing we do know is we are going where we can afford to fail.

As Benjamin Franklin once said, failing to prepare is preparing to fail.

Mr. SCHROCK. That's right.

Mr. CILLUFFO. So I think we need to identify some of those areas that maximize secondary and tertiary benefits beyond just guards, guns, and gates, and training and exercising, getting people in the same room together at the highest level and at the operating level will go a long way in at least breeding some of that trust, because ultimately that's the word.

It's not that people distrust one another. It's that they don't appreciate their roles and their missions and I think it takes time and we'll need to reach out to the American people to garner their trust and enlist their trust.

Mr. SCHROCK. Mr. Chairman, thank you.

I could stay here all day but I have another appointment I must go to, but I thank you very much.

It has been very, very beneficial to you being here.

Mr. SHAYS. Thank you, Mr. Schrock.

Mr. Tierney.

Mr. TIERNEY. I don't know if I heard Dr. Kass and Mr. Schrock correctly but there is a discussion about campaign discussions and what people are going to do about this and I don't know if I heard the correct statement, do you think this is healthy or not, and, if that's the case, I think it is absolutely healthy that we have a transparent discussion.

I think we all ought to be focused on the issue of terrorism and that we all want to deal with it but I think how we deal with it is essential.

To have a transparent discussion among all the candidates, as well as the incumbent, what is our approach to national security, what are our strategies going to be?

Mr. SHAYS. Will you yield?

Mr. TIERNEY. Sure.

Mr. SHAYS. What I heard was ultimately we have to have a national agreement and that we've never had the kind of debate that you're suggesting, that we just kind of——

Mr. SCHROCK. That's right. That's right.

Mr. TIERNEY. Then we all agree debate is important and critical.

Mr. SCHROCK. Administration after administration.

Mr. SHAYS. But it has to be dealt with on a bipartisan basis.

Mr. TIERNEY. Exactly.

I'm much assured to hear that because that's not something we've had so far and we've had a lot of politicking and posturing and setting things out without consulting the other party; sometimes without consulting Congress.

This committee is as frustrated as anybody as far as setting standards for our local communities, etc., in terms of what has not been done, in terms of looking at the local resources, and I think we have to know what people are going to do in that regard, what their attitude is toward this whole situation, and that may need to be clarified, I think.

All the things that the members of the panel have been talking about here in terms of coordinating, I assume you will agree it is just as important to coordinate the resources between the national and the international level; there would be no disagreement there, right?

Dr. KASS. Yes.

Colonel LARSEN. Yes.

Mr. CILLUFFO. Yes.

Mr. TIERNEY. I don't know there is a lot to ask in terms of questions, so, Mr. Chairman, I will yield back to you at this point.

Mr. SHAYS. Thank you.

Mr. Ruppersberger.

Mr. RUPPERSBERGER. Yes.

We're talking about plans so that we can get to our end game strategy for implementation, and, just as an illustration, just to have your opinion, we were very successful in the beginning stages of the war with Iraq, and then after we were in it all of a sudden we had problems, and there has been allegations that the planning for the post-invasion was not adequate, it was put together hastily, and it took a while to get to stabilize, to be able to bring the security that is needed to liberate Iraq.

Do you have any opinions about that plan and how it would relate to what we're talking about here today?

Dr. MCINTYRE. I've heard that discussion. I think it casts the question too narrowly. I have a problem not just with the issue of Iraq. I have a problem with the direction of military thinking since the end of the cold war, and it seems to me that, regardless of party, regardless of ideological background, regardless of service affiliation, there has been a relentless tendency over the last 15 years to focus on how we're going to do something instead of what effect of whatever it is we're going to do will have on the enemy.

I think we perhaps got off on the wrong foot after the end of the cold war, in 1989 to 1990 and 1991, when instead of asking the question how do you defeat enemies, why do people quit, why are wars over and then begin to construct our military to be flexible enough to achieve that, instead we focused on the question of how do we take new information technology and apply it to what we are doing to make it better.

That happens to cross-administrations from both parties, it's happened with conservatives and liberals.

I'm telling you I think we have not asked as a government, as a Nation, in the academic communities, in the service colleges, we have not asked the single most important, most fundamental question: Why are wars over?

We have focused instead on why wars start and if you ask that question why is war over and why do wars end then it takes you to a different pattern. You buy different things. You have a different set of planning, so I guess what I'm trying to tell you, sir, is I understand the criticism with how this war was waged.

My criticism, however, is much larger, and that is how all of us have been thinking about wars since trying to recast ourselves and our military for the last 15 years, and if you will take that different approach, I'd suggest the same approach to intelligence.

We keep asking the question: "What do we want to do?" The central question is: "What do we want the enemy to do?" That determines what we do.

Mr. SHAYS. What does that mean? Can you answer? What does that mean?

Dr. MCINTYRE. What do we want him to do, do we want him to surrender, to cooperate with us, want him involved along certain borders, to simply die, change his ideology?

What is it we want the enemy to do, and until we can figure that out, our applying different means is not going to solve the problem.

We are getting better and better with making the military more flexible, making the arrival of bombs more precise, the employment of forces more rapid. I'm not sure that solves the problem but just getting better at what we do.

Mr. RUPPERSBERGER. We are still considered to be the superpower of the world because of our technology but there was an issue that I believe occurred under Carter, Stansville Turner, where there was a policy decision made to take more away from human intelligence and to put it into the technological end, and, as a result of that, if you want to look at the whole picture that's happened right now, we do not have—we had it but we don't have it to the degree we need to have the human intelligence, that we know the culture of the people we're dealing with; I mean, just Iraq, we have religious issues that are out there. We have a lot of issues that we have to address and still—we still have to make sure we secure the area and that we finish what we started.

Dr. MCINTYRE. That's precisely what I was saying.

Mr. RUPPERSBERGER. To balance terrorism, and if you look at DOD, it's a huge massive agency, and the culture there was to go after as we did in the beginning of the Iraqi war and we were successful, but we also are dealing with terrorism now and it's a different ball game.

Dr. KASS. So it's just another aspect of education which most people forget, and that is the total lack of language skills, understanding of other cultures.

If you looked during the cold war, Congress legislated the National Defense Language Act. A lot of us who learned Russian during the cold war, myself included, benefited from scholarships which were designed to learn about our enemy.

We do not have that. We do not understand the enemy that we are fighting, and I would submit to you that is a critical step.

One of the problems in Iraq is not lack of planning, but it is lack of basic understanding of what the enemy might do, and you've got to be able to understand what he might do, based on understanding his culture, his history, his past behavior.

We don't have that.

Mr. RUPPERSBERGER. I agree with you.

We also need to learn more as a country about the Muslim religion.

Dr. KASS. Yes, sir.

Mr. RUPPERSBERGER. Because if we're ever perceived by Muslims there is a war against Islam, we'll have a very real problem and it got real close in the beginning of the Iraqi war, Egypt and other areas, but I think it turned around.

One other area I'd like to get into, you mentioned the issue of intelligence. Are you familiar with the Office of Special Plans in the Department of Defense?

Almost everyone involved there was more of a political appointment instead of a long time member of CIS, NSA, whatever, and there were concerns about that group circumventing, say, the CIA and not vetting all the information before it actually went to the policy of the President, and as a result of that there was actually information that really got into the State of the Union last year.

Do you think that there needs to be, when you have an Office of Special Plans, that there needs to be more of a relationship with that type of group and with the other agencies, such as CIA, NSA, FBI, that type of thing?

Dr. MCINTYRE. I think it's really important, sir, when you're being called upon to testify to your expertise, to know when to draw the line, and I don't have an expertise in that area, so any answer I give you would not be an expert answer.

I just don't have the expertise to answer that question.

Mr. RUPPERSBERGER. Do you have an opinion?

Dr. MCINTYRE. My opinion, sir, is that we have missed something much bigger than people are digging at right now. It was just not the Iraqi war we missed. We missed the response of the French, we missed the response of the Turks, we missed the response of the Russians.

We missed the way Saddam was going to play his hand and we missed it for a long period of time. We didn't get what was going on, so that is structural and is not specific to either this administration or the past one. It's a much larger conceptual problem, cause and effect.

Mr. RUPPERSBERGER. Why didn't we do that?

Dr. MCINTYRE. Well, what we taught at the National War College, if you're not real careful believing is seeing, and over a period of about 15 years, we built up, I think, a habit of we thought we knew what we were seeing and consequently we saw it, not just in this area, but in other areas as well and it is very hard to break that.

It takes outside thinking, outside expertise, a constant challenging, so I want to be very careful.

You asked me for an opinion. I can give you expertise as a strategist and I can tell you the history is filled with people who saw what they believed and you have to be careful about that.

I cannot judge this particular office. I just don't know.

Mr. RUPPERSBERGER. In your testimony, it was General Hooker, correct?

Dr. KASS. Yes, sir. Yes, sir, and that's a prime example, sir, just to reinforce: the notion of understanding your enemy and understanding your allies and not expecting others to behave the way you would in similar circumstances. We are not very good at it.

Mr. RUPPERSBERGER. One other issue. One other question?

Mr. SHAYS. Oh, no. Keep going. Keep going. It's fascinating.

Mr. RUPPERSBERGER. One other question in a different arena.

Your testimony, I forgot whose it was I read, talked about the issue of preemption, without bringing the rest of the world into the fold.

What do you think the administration could have done to bring the other nations into the fold before we did the preemptive strike with Iraq from a planning perspective?

Mr. CILLUFFO. Well, I'll take not Iraq specifically but looking at preemption and the war against terrorism and non-State actors, which actually requires personalizing.

When we deal with States, you need the information that exactly you would mention, and largely that's going to be based upon human intelligence and these people were not Boy Scouts, these aren't good people, and obviously good people don't have the insights into the mind of the terrorist, but, ultimately, from a preemption standpoint, obviously you want to bring along as many supporters as you have and we have on the war on terrorism.

We're working hand and glove and especially with respect to the indigenous security services. With many nations, we are not on a first name basis with them and good relationships with them, but with the war on terrorism we actually have been able to cooperate and coordinate with the foreign services and many—and I'm not speaking Iraq specifically but it does require making some hard decisions.

You've got to be willing to make mistakes. People have to be willing. Analysts aren't clairvoyant. They're going to make mistakes as well. If we were analysts, obviously, we would want to be on Wall Street and identifying where stocks are going in the future. It's an imperfect business, and all too often if people go out on a limb and they get caught for getting something wrong, they don't necessarily see the light of day in the future. So I think that both in the collection side, where people need to be willing to take risk and we need to accept some blowback and on the analytical side we need to be willing to make mistakes, and that is something that ironically is not fostered, to some extent, something I think that the Congress, in conjunction with the administration, can help play.

Colonel LARSEN. Sir, in line with what we've been discussing here, many of our allies understood the situation better than we did—what the end state would look like. We don't speak the language. There were 40 fluent Arabic linguists in the State Department when the Iraq war started, that's all.

Mr. RUPPERSBERGER. Do you feel that is a breakdown in our intelligence then?

Colonel LARSEN. It's bigger than intelligence, I'll agree with Dr. Kass.

Dr. KASS. It's in the nation.

Colonel LARSEN. We're talking about State Department, Department of Defense, it's national security that we don't understand who we are at war with, which goes back to Mr. Clausewitz's first statement, you better understand what you're getting involved in.

Mr. RUPPERSBERGER. It's not only in human intelligence but in analysts. We have to connect the dots.

Colonel LARSEN. Analysts and policymakers.

Mr. RUPPERSBERGER. But eventually the policymakers are relying on the intelligence to make their decisions, and that's why it seemed to me there was a circumvention of a standard that was used in the past that wasn't used, and what are we here about?

We're here to learn about what we did wrong, so that we can fix it and make it better. Bottom line, that's where we want to go.

Dr. MCINTYRE. Let me give you two brief points on preemption, since it is such an important topic.

This is actually what I did my dissertation on about 5 years ago, modernization forces, and I came to the conclusion in 1999 that the United States was going to be moving inevitably toward a doctrine of preemption during the last administration because that's just where the logic of war takes us.

I concluded in looking at previous wars that there were two things that caused a Nation to preemption, to attack preemptively. One is if it decides that the threat against is so overwhelming that it won't be able to survive the first strike, then it will have to preempt.

The second is, alternatively, if it decides that its own capabilities are advancing to the point that a strike would be relatively easy and relatively low cost.

What we had in the Iraq war was the perfect storm. Both of those things came together. We had a situation where we had every reason to believe that an attack against us, for example, of biological weapons, would be a one-blow knockout. No. 2, we had every reason to believe we could take care of this relatively quickly and with low cost and I guess what that tells me is that we need to be really, really careful because the momentum for any administration will be to be pulled forward by such circumstances.

Mr. RUPPERSBERGER. Thank you.

Mr. CILLUFFO. Mr. Congressman, and your intelligence should support decisionmakers, that's key. It's not the decisionmaker itself. That's something that's underappreciated or misunderstood.

Mr. RUPPERSBERGER. OK. Thanks.

Mr. SHAYS. This has been a fascinating panel, and the questions asked. I feel in some way like I'm losing track of the original effort of our committee.

Mr. RUPPERSBERGER. All esoteric.

Mr. SHAYS. If you ran against an opponent, ultimately you would want your opponent to lose. You would then maybe bring it up one level and say you would like to get out of the race before you lose, and third would be you would like them to actually endorse you.

Mr. CILLUFFO. Right.

Mr. SHAYS. I mean, I would love al-Qaeda to just love us and the world would be peaceful. I know that's not going to happen.

I am fascinated though by certain concepts. I've been to Iraq four times, one time just without the military entirely, two times without the military and then with the military and one time just with the military, so four times total, and there was one individual named Mohammed Abdul Hassan and he grabbed me by the arms, by the shoulders practically, and he said you don't know us and we don't know you.

That was in April, and I just came back to our folks. We've got to get our Arabic speakers in there and Iraqi Americans as fast as we could.

Now, what's surprised me, Dr. Kass and Dr. McIntyre, is I put the blame on this squarely on the military and the White House, because I agree with Mr. Ruppersberger. We went down in April, May, and June, and July and we've been clawing our way up since August, and we've made some progress, so if in Iraq we were here in April and we're here now in February, there's some slight incline. It's more significant because we got ourselves deep in a hole.

How in the world, though, given what you all teach, which I totally accept, how would we have blown it? Why would the military have been the one to have blown it in that sense; or let me say this: Was it the military saying in your judgment we better be careful, and it was maybe the political leaders not listening to the military?

I know this is a little sensitive, but this is big stuff for me.

Dr. MCINTYRE. Sir, we'll go wherever the chairman wants to go in the discussion. I don't place the blame for this on the administration. I do not place it on the political leaders, and I do not place it on the military. I place it on the academic community.

We have been thinking about the wrong things for 40 years. It is not just the intelligence community that was caught totally by surprise by September 11. It was the academic base from which the intelligence community is drawn. That's who educates our people.

Dr. KASS. Yes. Yes.

Mr. SHAYS. But you were both there and you're persuasive; I mean, I wouldn't have been in your class and this not been memorable if you had discussed these things.

Dr. MCINTYRE. At the military colleges where Randy taught, they draw what they teach from the civilian academic community, and so there is a limited amount of discussion to draw from, and I'm just telling you, sir, since 1950 or 1960 we haven't talked about how to end wars in the academic community. We talked about how to prevent them, so there is a very limited body of knowledge out there to draw on.

Colonel LARSEN. Let me give you a very specific example that answers your question.

Strategically, I think many of us agree Saddam had to be done away with, perhaps establishing a democracy that's going to make the world safer. Tactically, our troops did a marvelous job. Operationally is where I saw some failures. One hard example and this is from Lieutenant General Paul Surgeon, who is retired and in charge of rebuilding the entire Iraqi Army.

It was a great plan. Unfortunately it ended up like General Hooker's because the troops were supposed to lay down their arms at the barracks, in place, stay in uniform, we would take them over, so we had a bunch of good Iraqis that had some bad leaders and now we have a police force and military that we can quickly put leaders upon.

CENTCOM Headquarters when the war started a couple days early said lay down your arms and go home. They threw off their uniforms and went home and blended back into society. We don't know where they were, so our whole plan for controlling the country afterwards fell apart at the operational level.

Mr. SHAYS. Well, there was a big discussion because they were—I mean, not big distraction, but we could spend a lot of time here, because, for me, Dr. Kass, you started out not the Hooker part but the humility part was what caught me.

Dr. KASS. Yes, sir.

Mr. SHAYS. Because given what I thought and the President thought and the French even thought and the Germans even thought, that we would find weapons of mass destruction. There were a few Members of Congress who didn't think that and I acknowledge that, but it strikes me that a little less hubris is in order.

Dr. KASS. Yes, sir.

Mr. SHAYS. And what struck me is that there was just tremendous arrogance having won this war, even without the Turks' help, because that was a whole theatre we weren't able to enter in and we still did it.

It's hard not to feel like, boy, things are going well and then it just kind of fell apart for a few months, and hubris is the thing that I put in the biggest challenge.

Yes, what were you going to say?

Dr. KASS. Yes, I couldn't agree more with you. We are the victims of our own success. Being the world's superpower, having our products, our music, our entertainment spread globally makes us believe everybody likes us.

They don't. They don't want to be just like us, but we somehow fail to understand that.

You asked, couple of minutes ago, sir, why don't we understand the adversary?

The simple answer is: We don't study them. We apply our own modes of behavior, our own standards of rationality to the universe, and that is why we are quite often incorrect in our assessments.

Dr. McIntyre is exactly right. It comes down from inside, what we teach in our universities, in our colleges.

We are still wedded to the cold war paradigm of what we teach. That horizontal integration that we all talked about needs to be taught to our kids in high school and in college. It is not. By the time they become general——

Mr. SHAYS. I get your point.

It gets me, and I'm looking at Mr. Yim and I think he's probably thinking, what does this have to do with what we talked about; but I'm going to ask you to tie it up, Mr. Yim, or Doctor, because it gets to what John Tierney and I talked about.

As soon as we start reorganizing, we developed a national strategy, and one of the tragedies I think has taken place, tragedy is a strong word, but we have never fully had a dialog about what the threat is. So, for instance, I believe strongly, in the Patriot Act, not some of the other losses of authority by the general public and civil protections, but the Patriot Act I believe in strongly, and a lot of people don't in my district, don't because they don't think there's a threat. They honestly don't think there is a threat because we have stopped talking about what the threat is, and why we need it and that people, when our intelligence community had better intelligence and blame them, that they in my judgment don't want them to have a very important tool to get intelligence.

I realize we can all look at this differently but this is the kind of thing I'm sorting out. I'm thinking I hope that the Democratic candidate forces a dialog on this whole issue.

You know, what is the threat and how are we responding, and maybe in the end we are all going to come to an agreement that we all need to do all the things that we've done, but at least we'll all be in agreement. I don't know what you ultimately decide.

Let me ask this: What happens in the end if we can't agree on a strategy; in other words, one of the arguments is maybe we can't debate the strategy because maybe we can't agree to it. I mean, one of the important elements is there has to be a buy off, I think, with the general public, so maybe you can talk about that.

What happens if the public doesn't agree on a strategy? Should I assume that ultimately we can, we should do it, or should I assume that if we can't, something happens? What happens?

Dr. KASS. So let me take a stab at it.

Passion is good. Consensus is not necessary. I would submit to you that we have shied away from even identifying the enemy.

If you read the strategy skillfully, they tell you who the enemy is not. The strategies will tell you we are not at war with Islam, and so you mentioned that, but the strategies do not tell you positively who the enemy is or what the enemy is.

That is where you need to begin to build consensus. That is too fundamental an issue to skirt or void and jump immediately to. This is what I'm going to do about this.

This is another example of ready, fire, aim.

Mr. SHAYS. Anybody else?

Dr. MCINTYRE. Based on the discussion I had previously with Colonel Larsen, if we don't get a consensus bureaucracy takes control of the administrative part of this government.

Mr. SHAYS. If you don't consensus.

Dr. MCINTYRE. If you don't get a consensus on the strategy, the bureaucracy takes control of the future and local interests take control of Congress and the bureaucracy determines what we do and the local interests determine what we buy, and we find ourselves in a significant problem 10, 12, 15 years down the road, because those are the two things that will seize control.

Mr. SHAYS. Bureaucracy and what?

Dr. MCINTYRE. Bureaucracy and local interests. It's more the case that the people in your district will want certain types of spending in your district, so that's not exactly special to say I want you to take care of me in our district.

Mr. SHAYS. Can I put it in my words: They may not know what they need because there's nothing, so they just think—in other words, I want to understand this a little better: Are you suggesting that without some consensus or without a national strategy that everybody buys in, we go in a lot of different ways?

I don't understand.

Dr. MCINTYRE. You have to have a fire shield, I think, as a representative, in the same way that the only way we were ever able to close bases is if we were able to establish a set of priorities and rank the bases and then say local Congressmen can't be blamed by the fact that you didn't meet this priority.

You see, we've built a fire shield. I think we have to have some system of priorities to help build a fire shield for you and for other Members of Congress; otherwise the pressures will be to continue spending at local levels regardless of priorities. So two things will happen. I think bureaucracy will run things at the top and local requests will overcome and will be a constant strain on the budget.

Colonel LARSEN. I agree completely. The focus will be on Americans in your district as opposed to defending America. That's the sound bite for you. I agree completely.

Mr. CILLUFFO. Mr. Chairman, I think we do have some of the overarching strategies in place. It needs to be an execution and implementation and, as the old military adage goes, amateurs talk strategy, professionals talk tactics, they talk implementation and execution.

That said, I think your point, in terms of raising it and in terms of a debate and dialog, is absolutely crucial. We need to enlist and marshal and mobilize everyone in our generation's war against terrorism.

I spent a lot of time speaking, I've got four young daughters of my own and spend a lot of time speaking in public schools and other schools, and how do you send that message, while at the same time having it not become a self-fulfilling prophecy and creating fear. So I don't think we even had full consensus on a containment policy, so consensus shouldn't be the goal.

Mr. SHAYS. We did during the cold war, correct?

Mr. CILLUFFO. Not completely.

Mr. SHAYS. We may not have gotten a consensus on whether we need a missile defense system or something, but generally it was containment, reactive, mutually assured destruction.

Mr. CILLUFFO. For the most part, but it took a while to get to that point, and even at the end-state some would argue we were too hard in areas and not hard enough in others. It took a couple of key people who bridged, Scoop Jackson and a couple of others, parties to help mobilize the thinking along those lines, but I don't think we need to look for consensus, but I do think there are different actions that different constituencies need.

I don't want the general public being all that afraid, so if they're not worried about something happening tomorrow, that's one thing. If those that are on the front line, those that are going to turn victims into patients, our first preventers and our first-responders, if they're lulled into a sense of complacency, then I've got problems, and the same can go in terms of the international issue. So this is a long term challenge.

I think it would be arrogant to think we know the answers today.

Mr. SHAYS. Mr. Ruppersberger, do you have questions?

Mr. RUPPERSBERGER. Yes. First I believe your issue of implementation is extremely important. As far as your issue of bureaucracy and the local level, our system of government is a representative democracy and what really works is a strong leader.

If the strong leader has the plan and sets the goal and then works on the consensus and works on getting the votes, the system usually works.

The best defense against a strong bureaucracy is a strong leader, and it's about leadership, and if you really look at the politics in this country now, in my opinion, why Republicans control the Senate, the House, and the Presidency is because I feel Americans feel Republicans are better at national security and probably feel for some reason, and I don't agree with it, that Republicans are more patriotic. But if you look at polling, as far as general issues of education and other issues, people like what the Democrats do, but I think that issue more than anything else is the leadership. The issue of national security and the patriotism is a strong issue. So I'm not as concerned about the bureaucracy, whichever party is the leader at the top who is setting the agenda. What I'm concerned about though is the plan and the information that is getting to the President or to the leader and where he's going or she's going to make the judgment on where they're going to go, how they're going to implement the plan.

You talked about it, Mr. Cilluffo, and I think that's where we need to look, and, if not, that's why I think the argument—I remember, I wasn't here—but term limits. I think term limits were extremely dangerous, because if you have term limits the bureaucracy controls.

Mr. CILLUFFO. Yes.

Dr. KASS. Sir, leadership is key and I totally appreciate you raising that issue.

What helps a leader is having a bold idea that can light up, galvanize, support, both domestically and internationally, and that is why I suggested the pretty bold idea of abolishment. Containment to me is too passive.

Mr. SHAYS. Before we break, I'd like Mr. Yim—for you to just make some comments on what we've been talking about.

Also welcome you all responding.

I feel in one way like we're getting totally distracted and equally so, because maybe it's an indication that we were talking about things rather than theory, so I can gravitate more to that than others can.

Also, I think we were talking about some, I think, really fascinating issues.

Tell me, put some perspective on what you've heard, and also I wanted you to tell me if it was a strategy that was not part of the seven that I saw, and also I would like you, this panel, before we leave, and I don't want to drag this on, but I'd like to ask about the list of strategies that we were talking about and whether they are just countless strategies or should be or shouldn't be.

Yes?

Mr. YIM. I think, as an overall perspective, Mr. Chairman, I actually am, perhaps because of my success or failure rate, I'm willing to accept less than 100 percent solution, because I very rarely in my life have been able to achieve 100 percent solution, and I think when we talk about this issue of consensus I don't think it's absolutely necessary in the sense that I don't believe we would ever have 100 percent consensus. I don't think we need 100 percent consensus.

When I was working with the military, I could never get the Navy and the Air Force and the Marines and the Army to agree, but for OSD there was some commonality in the debate and we have so far to go in improving the debate that even if we only got a 70 percent solution, I would be pleased with a 70 percent solution.

Ms. Kass talks about Clausewitz. There's another philosopher Goethe. I'll paraphrase it and destroy the quote. Just start something because when you start something there's a whole other bunch of events that come to play that you may not even have imagined once you started embarking on that path and things may have come to your assistance that you may not have anticipated.

I think for Homeland Security we are so far at the beginning that if we can arrive at a 60, 70 percent. solution——

Mr. SHAYS. Define what you mean by solution. We were talking strategies and standards and you're talking solutions. I'm confused by that term.

Mr. YIM. I'm talking an all-hazards approach, for example. If people are talking about we have to focus on bioterrorism as opposed to a bomb in the port or as opposed to agriterrorism, we have to buy this type of equipment versus this type of equipment. If you really look at those scenarios, let's look at the five high-risk scenarios, a bomb in the port, a bioterrorism, agriterrorism, a cyberattack, something like that.

If you really think about it, probably each of those different scenarios, even if in different jurisdictions of different agencies, they're probably about 60 to 70 percent of what you would do. The prevention and recovery is probably the same.

Why don't we do that stuff; and I don't think we focused enough on the common stuff that we can do. Other things are going to happen. There is going to be new technology. Nobody would have predicted the dramatic fall of the Soviet Union. I don't think we could have predicted that.

Things just happened, and I think that's really important for us. That means for me answering your second question what strategy are we missing?

I mean, I think we are focusing too much on Homeland Security Strategy. I think the strategy that we're really missing, Mr. Chairman, if I could be so presumptuous, this is not a GAO position, I've been increasingly concerned about the gap between the condition of our infrastructure and the ability or what we're going to be demanding of our infrastructure in the future, the capabilities of our infrastructure to meet 21st century challenges, and by infrastructure I mean not only bricks and mortar, but people, the skill sets, the education level of our people.

We are not devoting enough money to recapitalizing our infrastructure. We can talk something as simple as bridges. We all know that many bridges are deteriorating. They are not going to be able to handle the traffic load.

Talk about our hospital systems. They were not being recapitalized in a way that can handle SARS, a major league bioterrorism attack, and the gap is going to increasingly widen.

Mr. SHAYS. Let me see if I understand. So if we decide—it's fairly obvious that our electricity grid is just substandard, shouldn't that be part of a national strategy related to the war on terrorism or not?

Mr. YIM. Yes, but we are debating energy policy and security energy recapitalization. Those debates are to focus on certain things.

Why aren't we building in energy security, homeland security into fundamental decisions in recapping the power grid? We are talking. Why aren't we talking about making the transit systems more secure while we talk about recapping Amtrak?

Mr. SHAYS. Let me just go to the panel and go to Mr. Ruppersberger and then we're going to end here.

Any comments to be made?

Mr. CILLUFFO. I fully support Mr. Yim, especially the maximizing secondary and tertiary benefits to get a return on investment beyond just guards, guns, gates, and you can splice that so many ways, and the President in his budget for 2003 and 2004 actually did put a close eye toward achieving that; for example, enhancements in improving our biological warfare really is about epidemiological surveillance and disease surveillance, which was really a public health structure that was broke and broken, so there was attempts to try to maximize some of that.

I think we can go further, but I think one of the points here is that security for the American people is always too much until the day it's not enough, and that's something we need to keep in mind. It's not fun. It's not easy.

There are no ways to—defining success is a huge challenge, but I can tell you one thing I think the President and the Congress as well—and I honestly do appreciate in terms of the actions that were taken. We can't go to the American people and say what I coulda, shoulda, or woulda but didn't because of this or that. We need to act and act decisively.

Mr. SHAYS. I would go on forever, but we have a 1 o'clock closed-door briefing and I think with you, Mr.Yim.

Mr. YIM. With Mr. Decker.

Mr. SHAYS. With Mr. Decker, I'm sorry. We'll do that at 1. We need to end up.

Just any closing comments?

Mr. RUPPERSBERGER. Just infrastructure. I agree with you but infrastructure costs money. Gets back to leadership again. Leadership has to prioritize and if the economy isn't doing well, and I'm not, in any way, making this political. I mean, do you stay with a tax cut, do you stay with funding education, do you stay with all these different issues? So we know that infrastructure makes you stronger and it's probably pretty wise politically in the end, but it's the will to top, and again the decisionmaker, getting the advice on where to prioritize and put the money.

I can tell you this: If and when there is another incident like September 11, all of a sudden you will see reprioritization of money going back into homeland security, and in a way that's unfortunate but that's the way it's going to be, and if you could just comment on that.

We could go on forever. This is an enlightening panel, and, Mr. Yim, you've done a good job, and why we've gone off the subject matter is because we want to get to the bottom line.

Dr. KASS. Yes.

Mr. RUPPERSBERGER. Hopefully, we can learn from our mistakes and move forward.

Mr. YIM. I think one of the keys is long-term strategy. Even when we budget for recapitalization, we look at the value within the OMB scoring period, which is typically 2 to 5 years, and most of the value recapping an infrastructure occurs in the 10th year, something like that.

We have too short-term of a perspective I think in analyzing the strategies. The terrorists have 100, 500-year plans. We have 2-year plans.

Mr. RUPPERSBERGER. That's a culture, though.

Look at Scheiner vs. the United States. We want it now and we get it now. We're effective in doing it.

OK, thank you.

Mr. SHAYS. Yes, we get it.

Thank you all.

Mr. Tierney, any closing comments?

Thank you all for your participation. It's been very interesting. I appreciate it and I appreciate the indulgence of the audience here.

Thank you. This hearing is adjourned.

We will be having a closed-door briefing in room 2003 at 1 o'clock.

Thank you. Just to finish up. It will be a fairly short meeting, I think.

[Whereupon, at 12:49 p.m., the subcommittee was adjourned.]

○